Commission of the European Communities

INTENSIVE FARMING AND THE IMPACT ON THE ENVIRONMENT AND THE RURAL ECONOMY OF RESTRICTIONS ON THE USE OF CHEMICAL AND ANIMAL FERTILIZERS.

(Study prepared for the Directorate General for Agriculture)

Document

This document has been prepared for use within the Commission. It does not necessarily represent the Commission's official position.

D
628.1684
INT

Copyright ECSC-EEC-EAEC, Brussel - Luxembourg, 1989
Reproduction is authorized, except for commercial purposes, provided the source is ackowledged.

PREPARED BY

TEAGASC

The Agriculture and Food development Authority

IRELAND.

TABLE OF CONTENTS

Chapter		Page no.
	Executive Summary	I
0	Summary Report	V
1	Background to the Study	1
2	Situation Regarding Nitrates in Water in EC Member States	3
3	Intensity of Farming	8
4	Fertilizer Use Practices in the EC	19
5	The Nitrogen Cycle and Pollution	31
6	Model of Nitrogen Use	36
7	Estimates of the Impact of Restricting Nitrogen on Crop Production and Nitrate Concentration	58
8	Policies and Measures to Control Pollution in Member States	101
9	Principles and Policy Options	106

APPENDICES

0. Executive Summary

While acknowledging that the polluter pays principle is accepted, it is difficult to implement in relation to nitrate pollution. This is due to the difficulty in attributing responsibility for nitrate pollution among farmers, because of the diffuse source of pollution and the complex process by which nitrates reach the ground water. While the denitrification of water remains an expensive process, the policy strategy focusses on prevention of nitrate pollution from farming.

The first element of this strategy is to educate the public generally but especially farmers, about the impact of farming on the environment, particularly in relation to nitrate leaching and water quality. This requires the development of codes of good farming practices, designed to improve the efficiency of nitrogen uptake by plants and reduce nitrate leaching. Enlightened land use management and voluntary restraint could make a significant contribution towards reducing nitrate pollution from farming. Within this type of framework certain environmentally-friendly farm practices (such as ploughing-in straw to contain autumn nitrate leaching) could be promoted and ultimately adopted.

A tax on nitrogen is ineffective in controlling the level of nitrogen use. In order to reduce the optimum level of nitrogen use by 10 per cent, its price would have to be increased by between 50 and 100 per cent. Clearly, the optimum level of nitrogen is quite insensitive to prices. Since the rate of tax required to achieve a specific level of nitrogen use is very high, some system of tax refund to producers would be necessary. The revenue from a tax on nitrogen would be available to compensate producers, who could be given a refund of the tax for nitrogen used up to the specified limit. While such a tax approach would work in principle, it would require general monitoring of nitrogen use, even where its use did not give rise to environmental problems. The costs of administering tax collection and refunds would mean less funds available for tax refunds. Furthermore, it would be very difficult to enforce what would in effect be a two-tier price for nitrogen, as farmers could benefit by "illicit" trading between those who are entitled to a full tax refund and those who are not. A tax on nitrogen fertilisers is therefore an inappropriate mechanism for restricting nitrogen use.

Where protection of water quality required a severe reduction in nitrogen use, the economic incentive to ignore official limits would be strong and regulations or incentives would be required to encourage conformity with restricted nitrogen use. A tax on nitrogen to reduce its optimum rate of use seems infeasible. A farm quota for nitrogen seems problematic because the monitoring and direct control of nitrogen use on farms is rather difficult. An alternative is to have to indirect control by regulating land use and livestock intensity.

Regulating nitrogen use indirectly, through the cropping pattern and livestock intensity on the farms, would complement the implementation of codes of good farming practice. Farmers could be given the option of direct control of nitrogen use, rather than changing crop mix, where procedures for control of nitrogen use are agreed with the authority implementing controls on nitrate pollution. As regards nitrate pollution of groundwater, the relevant cropping pattern to control is that for a water catchment area. Excess nitrate leaching from some farms could be offset by low leaching from other farms in the same water catchment. Restrictions would only affect farms contributing excess nitrates and only to the extent necessary to attain the desired water quality for the water catchment as a whole.

Income losses arising from conformity with a restricted pattern of farming could be estimated by reference to the pattern of land use in a recent period. Compensation, over a transitional period, would apply only to those farmers suffering an income loss relative to income expected from their land used in the reference period. Community funds should make a contribution toward transitional compensation for loss of income arising from new restrictions and toward the costs of investment needed to implement the required farming practices. Regulations to control nitrate leaching from farming should be drawn up in consultation with the Commission if they are to qualify for Community funding.

Controls on either nitrogen use or land use in farming need to be designed for the particular water catchments where nitrate pollution of groundwater is a problem or is likely to become a problem. While health standards are set at Community level, the design of regulations to achieve these standards will have to be established locally and implemented at farm level by a local authority. The relationship between the implementing agency and individual farmers might be operated under a management agreement system whereby a farmer agrees to manage a farm in a prescribed way so as to achieve some desired environmental objectives. Management agreements are also an appropriate instrument for achieving wider objectives for the rural landscape and environment and for land use policy in general. They could provide a mechanism through which to integrate agricultural and environmental policies.

The problem of nitrate pollution of potable water has given rise to the necessity of monitoring the nitrate content of water supplies. There is also need for research to improve understanding of the nitrate leaching process and quantification of the relationship between nitrogen use on farms and nitrate pollution of groundwater. A tax on nitrogen used in farming, to fund this work, may be justified on the grounds that the work is necessary if nitrogen use, which is potentially polluting, is to be allowed.

INTENSIVE FARMING

AND THE IMPACT ON THE ENVIRONMENT AND THE RURAL ECONOMY

OF RESTRICTIONS ON THE USE OF CHEMICAL AND ANIMAL FERTILIZERS

Summary Report

1. Introduction

The Commission's concern to integrate environmental and agricultural policies, has assumed increasing importance in the evolution of the CAP. The Commission's 1985 Green Paper on the CAP noted that "In the last decades agriculture has undergone a technological revolution which has profoundly changed farming practices. There is growing concern about the effects of such changes on the environment". Among the environmental problems the Commission has identified are "Water quality problems in many areas of intensive agriculture. These include eutrophication, nitrate and pesticide pollution mainly as a result of misuse and/or overuse of chemicals, animal manures and other organic material".

This study addresses the issue of restricting the use of chemical fertilizers and animal manures so as to avoid nitrate pollution of groundwater. The objective is to be able to estimate, at farm level, the impact of restricted nitrogen use in farm production on farm income and on the contribution to the level of nitrates in water. The purpose is to maintain water quality through corrective action focussed on zones vulnerable to nitrate pollution.

2. Lack of Empirical Data

This short-term study is limited by readily available data and confined to 'state of the art' knowledge about how nitrogen use impacts on both farm output and nitrates in water. An attempt was made to get data for "vulnerable zones", where nitrates in drinking water are likely, on present trends, to be a problem. Though the response to a questionnaire issued to experts in Member States was limited, the data received for one region in Denmark and UK were analysed in the report to explore their implications for the control nitrate pollution.

3. Nitrate Leaching, Nitrogen Use and Crop Yield

While nitrate leaching is a complex process, a simple linear relationship may be appropriate when nitrate leaching is related to total nitrogen applied in excess of that harvested in the crop. This is illustrated in Figure 1, where nitrates in the drainage water (NO3) is related to applied nitrogen minus nitrogen taken up by spring barley in Germany (represented as NA in Figure 1[1]). Nitrogen applied includes chemical nitrogen and the chemical nitrogen equivalent of animal manures. Drainage water is the water which drains down through the soil, as distinct from that which runs off the land surface or evaporates into the air. It is the drainage water that feeds the deep groundwater. The parameters of the linear relationship in Figure 1 will vary with soil and climate. Its slope for clay soils will be less steep than for sandy soils, as the latter facilitate drainage by their greater porosity. Higher rainfall leads to more nitrate leaching, but the higher volume of drainage water will lead to more dilute concentrations of nitrates in groundwater.

---

[1] Figures 1 and 2 are adapted from de Haen, H. (1982) "Economic Aspects of Policies to Control Nitrate Contamination resulting from Agricultural Production", European Review of Agricultural Economics, Vol 9.

FIGURE 1: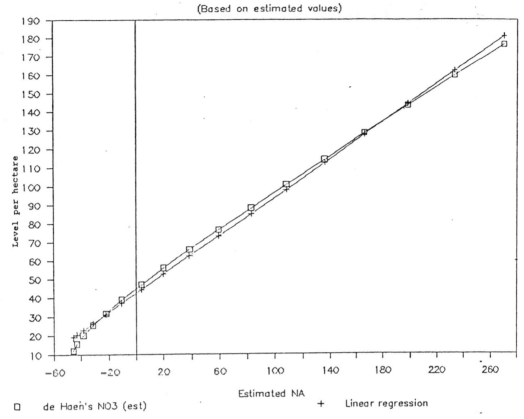

FIGURE 2: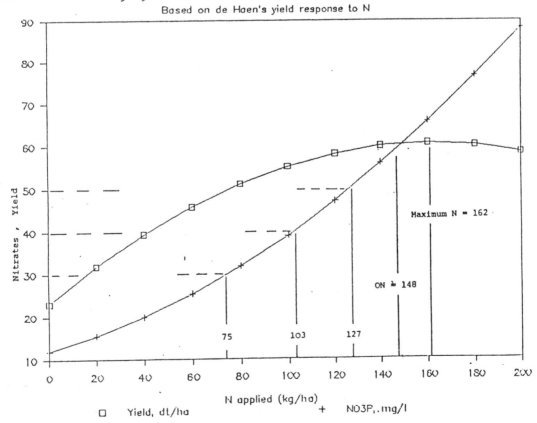

Applied nitrogen minus that taken up by the crop (i.e. NA in Figure 1) is negative when applied nitrogen is at zero. The crop is then depending totally on nitrogen released from the soil organic matter, which is a major contributor of nitrogen. Soils which are rich in organic matter and which facilitate release of its nitrogen, through mineralisation, will have higher nitrate leaching at zero nitrogen application.

The relationship between crop yield and nitrogen use is usually curvilinear, with the yield response decreasing as the rate of nitrogen applied increases. As nitrogen application rate is increased the proportion taken up by additional yield declines, so that nitrogen available for polluting drainage water increases at an increasing rate. Hence, as more nitrogen is applied, nitrate leaching increases at an increasing rate and a share of the nitrates leached out in the drainage water will be found to persist as nitrates in the groundwater. This is illustrated in Figure 2 for nitrogen use for spring barley production in Germany, assuming that 50 per cent of the nitrates in the drainage water is found to persist in the groundwater.

High nitrate leaching, at zero nitrogen application, arises from soil nitrogen which is mineralized, hence the importance of nitrogen released from the soil as a result of cultivation. This mineralisation occurs in late summer and autumn, mainly outside the period when nitrogen is being taken up by crops. The nitrogen released is leached out in the autumn and winter, particularly when there is no crop cover (as, for example, with spring cereal crops). Hence, the mineralised nitrogen contributes very little to crop growth and would be represented in Figure 2 by an upward shift in the curve representing nitrate concentration in the groundwater. In Figure 2 the current maximum allowable concentration (MAC) of 50 mg per litre is reached when nitrogen is 127 kg per hectare. If the intercept for nitrate concentration were 30 or 40 (as found by Pederson and Kolenbrander) instead of 12 mg NO_3/litre, then the 50 mg/l MAC would be reached by applying nitrogen at 80 and 47 kg/ha respectively. This highlights the importance of the intercept term, which represents nitrates from soil organic matter. Soil organic matter is the main source of leached nitrates under arable farming and its level is largely determined by the history of land use.

The extent to which nitrates leached persist as nitrates in groundwater can vary widely, in the range 30 to 80 per cent. In the illustrative calculations employed in Figure 2, a 50 per cent persistence rate was assumed. Persistence is higher in porous sandy soils and lower in dense clay soils, where slower percolation and anaerobic conditions facilitate the breakdown of nitrates, (denitrification).

4. Farmers' Income Losses from Restricted Nitrogen Use

The purpose of restricting nitrogen use is to ensure that nitrates in groundwater do not exceed the MAC. The guideline target is to have nitrate concentration at half the MAC (i.e 25 mg NO^3/litre) or less. A key question is: how would farm income be affected if nitrogen use were restricted so as to ensure that nitrate concentration in groundwater did not exceed the MAC?

First we need to know what limit to set on nitrogen use, so that nitrate concentration will not exceed the specified limit. This requires information on soil, climate, land use and farming practices for the groundwater catchment in question. Taking the example of West German data for a barley crop, illustrated in Figure 2, the nitrogen limit is 127 kg/ha for MAC at 50 mg/l. If the MAC were set lower, at say 40 and 30 mg/l, then nitrogen use would have to be restricted to 103 and 75 kg/ha respectively.

How such restrictions on nitrogen use affect the profitability of barley production depends on the extent to which the limit to nitrogen use is below the optimum level. Potential loss of income is the difference between income at the limited nitrogen use and the income that could be achieved if nitrogen were used at its economic optimum level. This optimum is found by increasing nitrogen application until the value of additional yield declines to equal the cost of additional nitrogen. Using 1986 prices, for example, the optimum rate for nitrogen in Figure 2 was 148 kg/ha. The appropriate limit to nitrogen use is affected by the level of the MAC, nitrate concentration at zero nitrogen (i.e. the intercept for NO3P) and the persistence of leached nitrates. Estimates of potential income loss for different values of these parameters are shown in Table 1.

Table 1: Potential loss in farm income based on Figure 1 for different MACs, intercepts* and nitrate persistence rates.

@ 1986 Prices: optimum nitrogen = 148 kg/ha

Persistence	Intercept (mg NO_3/litre)	MAC	N Limit (kg N/ha)	Potential loss in farm income ecu/ha	%
60%	12	50	106	43	18
50%	12	50	127	12	5
		40	103	52	22
		30	75	136	57
	30	50	80	117	49

* Intercept for nitrate concentration indicates its level when nitrogen application is zero.

At 50 per cent persistence, a MAC of 50 mg/l would have required nitrogen to be reduced below its optimum level to 127 kg N/ha, giving a potential income loss of 12 ECU/ha. The lower nitrogen limits to satisfy a MAC of 40 and 30 would give a potential loss of 52 and 136 ECU/ha respectively. So the potential loss increases steeply as the MAC is lowered, because the yield response to nitrogen is higher at lower rates of application. Hence, farmers have a vital interest in the level of MAC.

Soils with high nitrogen content, especially if cultivated, will have higher nitrate leaching and hence require lower restrictions on nitrogen use. If, for example, the nitrate concentration at zero nitrogen in Figure 1 were 30 instead of 12 mg/litre, the potential loss of income with MAC at 50 would be 117 ECU/ha instead of 12 ECU/ha. A higher persistence of nitrates leached in the drainage water would also require lower limits on nitrogen use, to satisfy the MAC. If, for example, the persistence rate were 60 as opposed to 50 per cent, this would have reduced the nitrogen limit from 127 to 106 kg/ha and increased the potential income loss from 12 to 43 ECU/ha. Clearly the required restriction on nitrogen use, and the associated income loss, are very sensitive to the parameters of the relationship between nitrate concentration and nitrogen application. These parameters vary between locations depending on soil/climatic conditions and the history of land use.

The potential loss in farm income can be quite significant. The estimates above range from 12 up to 117 ECU/ha, which would be about five to 50 per cent of the income on owned land. In some arable areas in the east of England, where nitrate concentration is close to twice the MAC, reducing nitrogen application rates from 185 to 100 kg/ha would reduce net farm income by 30 per cent. As estimates of the technical relationships for zones vulnerable to nitrate pollution are not available, estimates of the likely losses in farm incomes, due to restricting nitrogen use, cannot be established. The current lack of technical information, due in part to the fact that nitrate pollution is a relatively recent concern, points up the need for research in this area.

5. Off-farm Economic Impact of Reduced Farm Incomes

A reduction in farm incomes, by reducing farm expenditure, will reduce incomes in non-farming sectors. The full income impact for a Member State would probably be in the range 1.7 to 2.3 times the impact on income from farming. A higher multiplier reflects a higher degree of processing of farm produce and purchases of farm inputs. The income multiplier for rural regions would be smaller, probably in the range 1.3 to 1.8.

Limits on nitrogen use would reduce farm production and, in the case of surplus products, would reduce expenditure on storage and export refunds. The extent of such savings would depend on the rate of export refunds. Hence the potential loss from a Community perspective is less than the potential loss in farm income.

6. A Tax on Nitrogen

One of the frequently discussed methods, for inducing lower nitrogen use, is to increase its price by imposing a tax on nitrogenous fertilisers. If these prices were higher, it would give a lower economic optimum level of nitrogen use. In order to reduce the optimum level of nitrogen by 10 per cent, its price would have to be increased by between 50 and 100 per cent. Clearly the optimum level of nitrogen is not very sensitive to prices. A tax on fertilises would have a much smaller impact on income than a price cut that would result in the same optimum rate of nitrogen use. This is because expenditure on nitrogen is only about 10 per cent of the value of cereal output.

Since the rate of tax required to achieve a specific limit to nitrogen use is very high, some system of tax refund to producers would be necessary. The revenue from a tax on nitrogen would be available to compensate producers, who could be given a refund of the tax for nitrogen used up to the specified limit. While such a taxation approach would work in principle, it would require general monitoring of nitrogen use, even where its use did not give rise to environmental problems. The costs of administering tax collection and refunds would mean less funds available for tax refunds. Furthermore, it would be very difficult to enforce a two-tier price for nitrogen, as farmers could benefit by "illicit" trading between those who are entitled to a full tax refund and those who are not. A tax on nitrogen fertilises is therefore an inappropriate mechanism for restricting nitrogen use.

7. Nitrogen from Animal Manures

7.1 Efficiency

In discussing the relationships of yield and nitrate leaching with total applied nitrogen, nitrogen from animal manures was converted to its chemical nitrogen equivalent. Nitrogen in animal manure is less efficient than that in chemical fertilizers. Part of the nitrogen, in the form of ammonium, can readily volatilize and is lost to the atmosphere. Part of the nitrogen is in the form of organic matter and is not readily available to plants.

Attempts have been made to establish a systematic basis for estimating the efficiency of manures from different animals, based on the mix of mineral and organic nitrogen. Estimates of a general level of efficiency are given in Table 2. Efficiency in practice is very variable.

Table 2: Estimates of a general level of efficiency index (%) for nitrogen from animal slurry

Slurry type	(% mineral N)	Arable land		Grassland		
				Zero grazed		Grazed
		Spring (%)	Autumn (%)	Spring (%)	Autumn (%)	(%)
Cattle slurry	(40%)	44	22	30	18	20
Pig slurry	(50%)	52	25	38	20	25
Poultry slurry	(70%)	66	29	53	24	35
Calf slurry	(80%)	73	31	60	26	40

Differences in volatilisation of nitrogen are a major source of variation in nitrogen efficiency. The level of efficiency can be increased by application methods which reduce volatilisation. Efforts to reduce pollution from animal manures will also tend to increase their effectiveness. Efficiency indices need to be established for specific manuring practices.

7.2 Nitrate leaching from animal manures

The mineral nitrogen component in animal manures contributes to both plant growth and leaching, in the same way as chemical nitrogen fertilizers, but that is not the case for the organic nitrogen component of animal manures. Organic nitrogen is only released following mineralisation. With spring applications, for example, much of the mineralisation occurs late in the growing season and consequently the uptake by plants of nitrogen from the organic source is less than for mineral nitrogen.

Cultivation stimulates breakdown of the soil organic matter and release of its nitrogen. Nitrogen lost through leaching consists primarily of soil nitrogen mineralised outside the growing season. The extent of this leaching depends on the length of the growing season, which increases for the following crops in the order: cereals, potatoes, beets and grass. Crops which have nitrogen uptake patterns that better match the release of nitrogen in the soil will give rise to lower nitrate leaching.

The contribution of organic nitrogen to leaching is higher than for mineral nitrogen. This is due to the relatively high nitrogen release in the late summer and autumn and the high rainfall in autumn and winter. On grassland, this problem does not arise except when manures are applied in the Autumn. Under arable cropping, however, animal manure contributes more to nitrate leaching than would be indicated by nitrogen efficiency for crop yield.

In attempting to limit nitrate leaching it is essential to limit total nitrogen, from both chemical and animal sources. Animal manures also contain potassium and phosphate. Pig and poultry manure are especially rich in phosphate, so that phosphate pollution has become a problem where use of these animal manures has saturated the soil with phosphate. In the Netherlands, it is the allowable limit to phosphate application which restricts the use of animal manures. However, animal manures contribute only part of total nitrogen, so that regulation of phosphate applications will not necessarily translate into control of nitrogen applications.

7.3 Limits on the production and utilisation of animal manures

It should be noted that even permanent grassland can give rise to nitrate pollution of groundwater if total nitrogen application rates exceeded the recommended rates. This could arise from heavy application of animal manures, if grassland were being used as a disposal site for excess animal manures. It is essential therefore to limit the density of livestock. The appropriate limit would need to be established taking account of the cropping mix in a water catchment and transport of animal manures into and out of the catchment. This might be implemented in vulnerable zones through a "farm manuring plan", as is being done in Denmark. The implications for stocking density will vary between zones.

Animal manures on arable crops need to be lower than on grassland, due to their lower uptake of phosphate and nitrogen. The share of total nitrogen coming from animal manures varies widely between crops, so that a lower limit is appropriate for cereals than for root crops. Tighter restrictions may be required for pig and poultry slurry, due to their higher phosphate content. If manure applications were restricted to the limit of phosphate uptake by arable crops, this would probably be adequate to protect against nitrogen pollution from animal manures also.

8. Increasing Permanent Pasture

Arable cropping is characterised by higher nitrate leaching, due mainly to greater mineralisation of organic matter in the soil and in applied animal manures. The leaching of mineralized nitrogen from arable cropping can be reduced by crop rotations which provide crop cover in the autumn and winter. However, lowest leaching losses would arise if animal manures were applied to grassland in the March to July period. It should be noted, however, that low mineralisation of manure organic matter in the first year means a build up of soil organic matter. Subsequent cultivation of soil with a high organic matter content would give rise to rapid release of this nitrogen reserve. If this were released by mineralisation in subsequent years, it would result in higher leaching. Hence, the permanence of grassland is important in relation to nitrate leaching.

Data on permanent grassland indicate nitrate concentrations in groundwater which are below the current guideline of 25 mg/l, where nitrogen application is at generally recommended rates. If the proportion of land devoted to permanent pasture (or other environmentally safe crops) in a given groundwater catchment were high enough, then there would be adequate dilution of the high nitrate concentrations arising from other crops receiving optimum nitrogen applications. Restrictions might be placed on cropping mix, so as to attain the desired water quality, assuming that nitrogen will be applied at the economically optimum rate. This is an alternative to controlling the use of nitrogen directly.

Nitrogen use could be controlled indirectly by controlling cropping pattern. The relevant cropping pattern to control is that for a water catchment area. Hence it would be possible for excess nitrate leaching from some farms to be offset by low leaching from other farms in the same water catchment area. Restrictions on cropping pattern would only apply to farms contributing excess nitrates and only to the extent necessary to attain the desired water quality for the water catchment as a whole. Land cropping pattern would be much easier to monitor and control than nitrogen use. The technical information requirements would be less under this approach, as nitrate concentrations need only be known for nitrogen used at the recommended rates and observations under such rates would be more accessible, especially in vulnerable zones where nitrate concentrations would be monitored.

It is pertinent to note that control of nitrate pollution is only one of many environmental aspects that are of interest. These include other pollutants, (for example phosphorous), and the preservation or creation of desirable ecological environments. Control of cropping pattern could also be an instrument for achieving these wider environmental objectives and could form the basis of a more comprehensive approach to land-use management.

Where crop changes are being considered it would be necessary to take account of differences in overheads, as well as in gross margins. This is particularly so where cash crops are being replaced by grassland or other forage crops for animal feeding.

9. Principles and Policy Options

9.1 Water supply options

The over-riding policy objective is to ensure adequately low nitrate concentration in drinking water. There are a number of ways by which this can be achieved. Where water supplied from groundwater exceeds the MAC, the drinking water could be denitrified or mixed with cleaner water to dilute the nitrate concentration or clean drinking water could be supplied in bottles. An alternative or complementary approach is to reduce the leaching of nitrates from farm land into the groundwater supply.

9.2 Polluter Pays Principle (PPP)

This principle is now enshrined in the Single European Act. However, it is difficult to implement in relation to nitrate pollution. This is due to the difficulty in attributing responsibility for nitrate pollution among farmers because of the diffuse source of pollution and the complex process by which nitrates reach the groundwater. Underlying the "Polluter pays principle" is the notion of a social contract between the citizen, in this case a farmer, and the wider society. A 'reasonable' social contract requires that farmers perceive the ill-effects of exceeding the MAC to be significant. While the medical evidence in favour of any particular limit is beyond the scope of this study, it is essential to recognise that its basis and acceptance is important to cooperation in enforcing limits.

9.3 A strategy for environmental co-operation

Much of the nitrate pollution from farming has arisen as an unintended and often unknown consequence of farming practices. The policy strategy which gave rise to this study is directed toward the prevention of nitrate pollution from farming. The first element of this strategy is to educate the public generally but especially farmers, about the impact of farming on the environment, particularly in relation to nitrate leaching and water quality. This requires the development of codes of good farming practices, designed to improve the efficiency of nitrogen uptake by plants and reduce nitrate leaching. Enlightened land use management and voluntary restraint could make a significant contribution toward reducing nitrate pollution from farming.

9.4 Regulatory measures and economic incentives

Where protection of water quality required a severe reduction in nitrogen use, the economic incentive to ignore official limits would be strong. In such circumstances, codes of good practice are likely to need support from appropriate regulatory and/or economic instruments, to provide adequate incentive for their effective implementation at farm level.

A tax on nitrogen would not give effective control of nitrogen use, as discussed above. A farm quota for chemical nitrogen would also be ineffective, as total nitrogen use (including animal manures) needs to be controlled. The monitoring and direct control of nitrogen use on farms is rather difficult, being more difficult than for phosphate. An alternative approach, discussed above, would be to have indirect control by regulating land use and livestock intensity. Farmers could be given the option of direct control of nitrogen use, rather than changing crop mix, where procedures for control of nitrogen use are agreed with the authority implementing controls on nitrate pollution. Analyses of soil for mineral nitrogen in October and of plants for nitrogen content may facilitate control of nitrogen use. Restrictions, whether on crop mix or nitrogen use, would only affect farms contributing excess nitrates and only to the extent necessary to attain the desired water quality for the water catchment as a whole.

9.5 Compensation for farmers

Where nitrogen use has to be restricted below the economic optimum level, to ensure acceptable water quality, the gain in water quality will lead to a loss in income from farming. Who should bear the losses arising from newly imposed restrictions? The main beneficiaries of the past 'freedom to pollute with nitrates' were farmers and suppliers of nitrogen fertilizers. In so far as farmers were unaware that they were contributing to pollution, then the pollution might be more appropriately viewed as unintentional or accidental. However the situation regarding responsibility changes, when people are made aware that their farming practices do cause a level of nitrate pollution that is unacceptable to society as a whole. In particular, if restrictions are set for farming practice, so as to avoid pollution, then it is clear what farmers can do without contributing to pollution. Farmers might reasonably be expected to operate within constraints relating to nitrogen and land use, which are legitimately introduced to meet health standards.

While accepting that farmers should conform to approved farming practices, should they bear the full costs of any associated loss of income or be fully compensated? To indefinitely compensate farmers for farming so as not to pollute water supplies, would be to imply that they had a right to continue polluting but were yielding up that right in exchange for compensation. However, it should be borne in mind that the introduction of constraints on farming practices would, by altering the profitability of farming, reduce the return on investments made before the environmental constraints were anticipated. It seems reasonable that farmers should be assisted for a time to adjust to any 'change in the rules' governing farming.

Regulations to control nitrate leaching from farming should be drawn up in consultation with the Commission if they are to qualify for Community funding. Proposed restrictions on crop mix, livestock density or nitrogen use would have to be based on the appropriate technical relationships[1]. Regions could be obliged to provide best estimates of these relationships, along with their empirical basis, so as to agree on controls with the Commission and become eligible for EC aid. It seems appropriate that Community funds would make a contribution toward transitional compensation for loss of income arising from new restrictions and toward the costs of investment needed to implement the required changes in farming.
Environmentally desirable changes in farming system and practices, which require investment might be made eligible for farm development aid.

---

[1] Some of the established models, such as GLEAMS, may be useful for this purpose.

Additional compensation for income loss due to environmental constraints could be justified over a transitional period. Savings on the disposal of surplus commodities, attributable to reduced nitrogen use, could be distributed to alleviate the income losses sustained, without necessarily increasing total public spending.

In some situations income loss in food production might be offset to a certain extent by adapting the farming system to provide environmental goods or services, (such as game habitats and ecologically or aesthetically desirable landscapes). In relation to limiting environmentally undesirable developments, EC Ministers for the Environment recently concluded that the structural funds constitute a privileged instrument in achieving such aims, as well as for the integration of the environmental dimension in the agricultural sector.

Where nitrogen use is controlled on specific crops, the potential income loss could be based on the income loss from reducing nitrogen use below its optimum level. Where control is by regulating the mix of crops, income loss could be estimated by reference to the pattern of land use in a recent period. Compensation, over a transitional period, would apply only to those farmers suffering an income loss relative to income expected from their land used in the reference period.

9.6 Implementing controls

While health standards are set at Community level, the design of regulations to achieve these standards will have to be established locally and implemented at farm level, by a local authority. The relationship between the implementing agency and individual farmers might be operated under a management agreement system, whereby a farmer agrees to manage a farm in a prescribed way so as to achieve some desired environmental objectives. Management agreements are also an appropriate instrument for achieving wider objectives for the rural landscape and environment and for land use policy in general. They could provide a mechanism through which to integrate agricultural and environmental policy. Farmers' acceptance of restrictions, or of penalties for pollution, might be enhanced if they were also eligible for rewards for publicly desirable 'environmental products' of their farming. Thus it may be easier to promote good water quality as part of a wider programme of environmental enhancement, which might have possibilities of rewarding environmentally-friendly farming.

9.7 Monitoring and research

There is a good level of general information on the process of nitrate leaching from farm land, which can guide the development of a code of good farming practice. However, there is a lack of information for vulnerable zones, indicating a need for research to quantify the relevant relationships for such areas. Very little research has been done on nitrate leaching on grazed grassland. The development and parameterisation of quantitative models seem an appropriate framework for this research. Simulation models are needed for testing alternative farming systems and the technique of multiple goal programming may be appropriate for exploring trade-offs between farming and environmental objectives. In relation to monitoring at farm level, analytical methods for estimating the nitrogen content of soil and crops grown seem relevant.

The problem of nitrate pollution of drinking water has given rise to the necessity of monitoring the nitrate content of water supplies. A levy on nitrogen used in farming would yield revenue to fund work on monitoring and research. Such a levy might be justified on the grounds that monitoring and research is necessary if nitrogen use, which is potentially polluting, is to be permitted. Such a levy might apply to purchases of chemical nitrogen and of nitrogen in animal feeds, as the latter will ultimately be found in animal manures.

Chapter 1

BACKGROUND TO THE STUDY

The EC Commission's concern to integrate environmental and agricultural policies, has assumed increasing importance in the evolution of the CAP[1]. They state that "objectives, such as self-sufficiency in foodstuffs or, socio-economic ones, such as maintenance of farmers' incomes have be complemented by objectives relating to the protection of certain traditional values which agriculture represents in our society, in particular with respect to the protection of the environment"[2]. This orientation has already been reflected in the modifications of Regulation 797/85 and was given emphasis in recent reviews of the CAP. The Green Paper of 1985 noted that "In the last decades, agriculture has undergone a technological revolution which has profoundly changed farming practices. There is growing concern about the effects of such changes on the environment"[3]. Among the environmental problems the Commission has identified are "Water quality problems in many areas of intensive agriculture..... These include eutrophication, nitrate and pesticide pollution, mainly as a result of misuse and/or overuse of chemicals, animal

---

[1] Commission of the European Communities (1988) <u>Environment and Agriculture</u>, Commission Communication, COM 338, June 8.

[2] ibid. p 1.

[3] Commission of the European Communities (1985) <u>Perspectives for the Common Agricultural Policy</u>, Commission Communication, COM 333.

manures and other organic material"[1]. This study addresses the issue of restricting the use of chemical fertilizers and animal manures so as to avoid nitrate pollution of groundwater.

The objective is to be able to estimate, for farm level production, the impact of restricted nitrogen use in farm production on farm income and on the contribution to the level of nitrates in water. In view of the relatively short study period, we have had to confine ourselves to readily available data and to 'state of the art' knowledge about how nitrogen use impacts on both farm output and nitrates in water. An attempt was made to get data for "vulnerable zones", (where nitrates in potable water are likely, on present trends, to be a problem) so as to focus impact analysis on these zones, where control of nitrogen is envisaged. There was a limited response to a questionnaire (see Appendix 1) seeking these data, which was circulated to delegates at the CEPFAR meeting (March 23-25, 1988) in Brussels and again at the end of April 1988. While analysis of available data was helpful in identifying appropriate interventions to control nitrate pollution, there is an obvious lack of data for areas where nitrate pollution is a problem.

---

[1] Commission of the European Communities (1988) op. cit. p 6.

Chapter 2

SITUATION IN EC MEMBER STATES REGARDING NITRATE LEVELS IN WATER

This section is based on submissions of EC Member States to a conference entitled "European Study Days in Water Pollution and Intensive Farming", CEPFAR, held in Brussels, March 23-25, 1988. An extensive summary of each Member State's submission relating to this section is contained in APPENDIX 2 of this report. Opinions or comments contained in the appendix summaries reflect those of the Member State delegations and do not necessarily represent the views of the study team. Furthermore, the data and reports presented at this conference have been taken to reflect the current situation with regard to the relevant issues in each Member State.

2.1 General nitrates situation in the EC

In general, the incidence of serious nitrate contamination in the European Community is relatively low. There is, however, tremendous variation in the incidence of excessive nitrate levels (i.e. above the EC Maximum Admissible Concentration - MAC - of 50 mg NO_3/litre of drinking water) between and within individual Member States.

The majority of Member States are concerned about the prospect of nitrate contamination in the coming years. The main contributory factors are generally considered to be related to practices associated with intensive agricultural systems involving high application rates of chemical nitrogenous fertilizers and/or the production of excessive volumes of animal manures resulting from intensive rearing and feeding regimes. Other contributory factors to nitrate contamination of either agricultural or non-agricultural origin were also acknowledged. In the main, however, known local or regional nitrate problems are attributed by national governments to nitrogen emanating from intensive agriculture.

The study team have been unable, during the study period, to obtain a full set of national data on recognised nitrate vulnerable-zones for each Member State. These data, which include soil and climatic conditions, land use and livestock numbers, recommended and actual fertilizer use and nitrate leaching indicators, are considered as essential in order to carry out meaningful economic analysis of the impact of restrictions on nitrogen use in individual vulnerable zones.

APPENDIX 1 shows these data requirements as a questionnaire which was transmitted to national competent authorities. Data in the form requested were submitted in respect of the Western Jutland region of Denmark and the Eastern region of the United Kingdom. The results of analysis conducted on these data are given in Chapter 7. In cases where non-agricultural sources of nitrate pollution have been identified, the data, where available, are often out of date, incomplete or selectively sampled. In some Member States, there is no systematic national sampling for nitrate levels in drinking water sources.

Generally, the concern is more pressing in the case of groundwater supplies than for surface water, particularly since the extent of future nitrate levels due to historical factors is still largely undetermined. Table 2.1 shows a recent estimate of the proportion of drinking water supply coming from groundwater sources in each Member State.

Table 2.1: Proportion of drinking water from groundwater sources

EC Member State	Percentage
Belgium	76
Denmark	99
Federal Republic of Germany	68
Greece	na
Spain	40
France	68
Ireland	20
Italy	88
Luxembourg	73
Netherlands	65
Portugal	70
United Kingdom	32

SOURCE: Reproduced from Worthington (1987)[1]

Those Member States dependent on groundwater for drinking water supplies would appear to be most concerned about the nitrate problem either at present or for the future. The relatively low groundwater dependency in the UK hides the fact that nitrate problems occur mostly in regions which are more dependent on groundwater than surface water.

---

[1] Worthington, P. (1987) "A Movement towards Community Measures for the Protection of Freshwaters against Pollution from Diffuse Sources of Nitrogen Compounds", Paper delivered to the Conference on Impact of Agriculture on Water Resources - Consequences and Perspectives, Berlin.

2.2 Nitrate problems in Member States

On the basis of submissions to the aforementioned conference, the current position with regard to nitrates in drinking water at national level are summarised hereunder.

Belgium

(1) French speaking region: MAC not exceeded,
                              Nitrate levels expected to increase.
(2) Flemish region: MAC not exceeded,
                              Concern about individual boreholes.

Denmark: Two per cent of groundwater above MAC,
                              Groundwater problem worst in Western region,
                              Surface water generally above MAC.

Federal Republic of Germany: Six per cent of drinking water above MAC,
                              Concern about expected higher nitrate levels.

Greece: No systematic monitoring for nitrates,
                              Some nitrate problems encountered.

Spain: MAC exceeded in some surface water,
                              Nitrate problems generally increasing.

France: Regional variation in nitrate levels attributed to agricultural intensity,
                              Concern about future nitrate levels.

Ireland: MAC not exceeded,
                              Problems with a few local boreholes.

Italy: No data available but some provinces have nitrate problems.

Luxembourg: MAC exceeded in some small district water supply areas.

Netherlands:	MAC exceeded due to agricultural activity, 25 per cent of drinking water will have nitrate problems in the future.

United Kingdom:	Two per cent of public water supply above MAC,
Nitrate levels expected to increase.

Chapter 3

INTENSITY OF FARMING

The increase in nitrogen loading, arising from farming, is due to increased chemical fertilizer use and purchased animal feedstuffs, which add nitrogen to the farm production cycle. The associated increase in crop yields and animal feed has resulted in more intensive livestock production. In certain regions, and in specific locations, the production of animal manures is high relative to the land available for its disposal.

3.1 Animal manure production

The production of nitrogen by animals, in manure, is based on the amount of feed required by the animals to attain their levels of production. Based on Dutch data, Sluijsmans[1] estimates annual production of N by an adult cow at 89 kg. This is consistent with an earlier report published by the Commission, which gave a figure of 90kg of N for a dairy cow and also gave the production of N from manures of other animals in terms of nitrogen cow equivalents[2]. We are using these cow equivalents, and a production of 90 kg of N per cow, to estimate the N produced by animals. It should be noted

---

[1] Sluijsmans, Ir.C.M.J. (1983). Final Draft Report on Practical Guidelines for the Farmer in the EC with respect to Utilisation of Animal Manures, Institute for Soil Fertility, Haren, Netherlands.

[2] Commission of the European Communities (1978b) The Spreading of Animal Excrement on Utilized Agricultural Areas of the Community: Summary and Conclusions, Information on Agriculture, No. 51.

that these are approximations, as the level of feeding to any given type of animal varies across the EC.

Table 3.1 gives factors for converting annual nitrogen produced by farm animals in slurry (urine and faeces) to adult cow equivalents. Applying these factors to livestock numbers in the regions of the EC and assuming 90kg of N per cow equivalent, gives an estimate of the regional production of nitrogen in animal manures.

Table 3.1: Factors for converting nitrogen produced by farm animals to adult cow equivalents.

Animal type	Nitrogen conversion factors (nitrogen cow equivalents)
Adult cows	1.0000
Calves (<=1 year or <=220 kg)	0.3000
Other bovines	0.6000
Brood sows >50 kg	0.2447
Other pigs - >20 kg	0.1330
- <=20 kg	0.0532
Sheep	0.1500
Goats	0.1500
Laying hens (100)	0.9090
Broilers (100)	0.4546
Other chickens (100)	0.1667
Geese (100)	0.7778
Ducks (100)	0.7778
Turkeys (100)	0.2000

Source: CEC (1978b) op. cit. Table 1.

The most recent livestock enumeration data available for EC regions relate to December 1985. Poultry data are not available and some livestock categories are less detailed than indicated in Table 3.1. Cattle other than cows are not differentiated by age and pigs are not differentiated by weight in the regional statistics. In order to estimate cow equivalents for pigs and cattle other than cows for each region, the numbers in the sub-categories for individual Member States concerned as at December 1985 (from Eurostat, 1988, Tables IE 1&2) were used to produce national weights using the conversion factors in Table 3.1 for the categories in question. These weighted conversion factors for each Member State are given in Table 3.2. Livestock data for December 1985 were not available for Spain or Portugal, so 1986 data were a used as weights in estimating cow equivalents for pigs and cattle other than cows. The relative stability of these weighted conversion factors across Member States suggests that using the State coefficient for each region within the State would give reasonable estimates of manure production from these categories of animals.

Table 3.2: Weighted conversion factors for nitrogen produced by pigs and other cattle to adult cow equivalents as at December 1985[*].

Member State	Weighted nitrogen conversion factors	
	Pigs	Cattle other than cows
Belgium	0.1252	0.4828
Denmark	0.1207	0.4187
Germany	0.1219	0.4361
Greece	0.1229	0.4045
Spain	0.1228	0.4259
France	0.1243	0.4676
Ireland	0.1256	0.4915
Italy	0.1270	0.4604
Luxembourg	0.1208	0.4815
Netherlands	0.1178	0.4298
Portugal	0.1233	NA
United Kingdom	0.1237	0.4680

* Data for December 1986 were used for Spain and Portugal.
Source: Eorostat (1988) Agriculture - Statistical Yearbook.

By applying the above-mentioned conversion factors to livestock numbers, various measures of the intensity of nitrogen production from animal manures can be calculated for individual regions for 1985. Data for grazing livestock were analysed separately for two reasons (see Appendix Table 3.1). Firstly, they are associated with land-based farm enterprises, as distinct from pigs and poultry. Secondly, manures from these animals are not used, to any great extent, on cereal crops and hence the cereal crop area is not available to absorb manure from grazing animals.

Data for horses and goats were not available for many of the regions, so nitrogen production intensity was estimated excluding these. Appendix Table 3.2 shows that the relative intensity across regions is similar, whether or not horses and goats are included. Data excluding goats and horses will be used in discussing the regional pattern of nitrogen production from animal manures, because of the better regional coverage of the data.

Examination of Appendix Table 3.1 shows that nitrogen from cattle and sheep manure had its highest intensity, relative to UAA (Utilised Agricultural Area), in the Antwerpen region of Belgium, with 250 kg N/ha. The intensity varies widely across regions. Most regions in the Netherlands and many in Belgium had over 125 kg N/ha, as did one region (Cantabria) in Spain. Excluding the cereal crop area increased the maximum N intensity slightly, from 250 to 258kg N/ha for the Antwerpen region while all the other regions also had an increase in intensity. However, for cereal growing regions, the exclusion of the area under cereals from the calculation gives a much greater increase in intensity. In general, the increase brings the intensity to over 125 kg of nitrogen per hectare in most regions of Belgium, in Luxembourg, in one region (Vest for Storebaelt) of Denmark, in many regions of West Germany, in Bretagne (France), in Lombardia (Italy), and in the West Midlands and North West regions of the United Kingdom.

Table 3.3 gives the estimated N production by cattle, sheep and pigs per ha. of utilised agricultural area (UAA). Data for poultry numbers are not available at regional level. The inclusion of pig manure in the calculation resulted in a maximum intensity of 403 kg N/ha. of UAA, occurring in the Noord Brabant region of the Netherlands. Intensity exceeding 200 kg N/ha was found only in some regions of the Netherlands and Belgium. According to Table 3.3, the regions with greatest nitrogen intensity from livestock production are Noord-Brabant, Gelderland, Utrecht, Limburg, Overijssel and Friesland in the Netherlands and Antwerpen, West-Vlaanderen, Oost-Vlaanderen and Limburg in Belgium.

Table 3.3: Nitrogen production by animals relative to Utilised Agricultural Area (UAA) and arable crops as a percentage of UAA

EC Member States and Regions	Nitrogen from cattle, sheep and pigs (kg N/ha. uaa)	Arable crops as percent. of uaa
BELGIQUE-BELGIE	180	53
VLAAMS GEWEST	251	57
REGION WALLONNE	123	50
BRUXELLES-BRUSSEL	71	60
ANTWERPEN	330	38
BRABANT	103	74
HAINAUT	116	64
LIEGE	146	38
LIMBURG	202	55
LUXEMBOURG	153	22
NAMUR	99	54
OOST-VLAANDEREN	257	56
WEST-VLAANDEREN	292	61
DANMARK	88	92
HOVEDSTADSREGIONEN	41	93
OST FOR STOREBAELT,EX.HOVEDST.	54	96
VEST FOR STOREBAELT	98	91
BR DEUTSCHLAND	98	60
SCHLESWIG-HOLSTEIN	103	55
HAMBURG	NA	52
NIEDERSACHSEN	101	60
BREMEN	NA	20
NORDRHEIN-WESTFALEN	115	68
HESSEN	83	66
RHEINLAND-PFALZ	62	59
BADEN-WUERTTEMBERG	87	55
BAYERN	104	60
SAARLAND	72	57
BERLIN (WEST)	66	71
ELLAS	34	51

Cont.

EC Member States and Regions	Nitrogen from cattle, sheep and pigs (kg N/ha. uaa)	Arable crops as percent. of uaa
ESPANA	25	57
NOROESTE	105	41
GALICIA	106	59
ASTURIAS	84	9
CANTABRIA	141	11
NORESTE	25	58
PAIS VASCO	71	42
NAVARRA	28	51
RIOJA	26	47
ARAGON	20	61
MADRID	24	53
CENTRO	19	64
CASTILLA - LEON	24	70
CASTILLA - LA MANCHA	12	64
EXTREMADURA	23	48
ESTE	38	41
CATALUNA	53	49
COMUNIDAD VALENCIANA	16	28
BALEARES	37	49
SUR	15	54
ANDALUCIA	14	53
MURCIA	25	61
CEUTA Y MELILLA	NA	NA
CANARIAS	13	64
FRANCE	55	57
ILE DE FRANCE	6	95
BASSIN PARISIEN	44	67
CHAMPAGNE-ARDENNE	30	73
PICARDIE	36	83
HAUTE-NORMANDIE	63	61
CENTRE	22	82
BASSE-NORMANDIE	89	37
BOURGOGNE	48	53
NORD - PAS-DE-CALAIS	67	72
EST	58	45
LORRAINE	59	49
ALSACE	53	66
FRANCHE-COMTE	60	32
OUEST	88	71
PAYS DE LA LOIRE	88	61
BRETAGNE	126	83
POITOU-CHARENTES	48	72

Cont.

EC Member States and Regions	Nitrogen from cattle, sheep and pigs (kg N/ha. uaa)	Arable crops as percent. of uaa
SUD-OUEST	56	53
AQUITAINE	50	55
MIDI-PYRENEES	49	59
LIMOUSIN	86	31
CENTRE-EST	55	35
RHONE-ALPES	46	38
AUVERGNE	67	32
MEDITERRANEE	16	21
LANGUEDOC-ROUSSILLON	15	22
PROVENCE-ALPES-COTE D'AZUR	18	27
CORSE	16	4
IRELAND	69	9
ITALIA	45	51
NORD OVEST	61	45
PIEMONTE	69	51
VALLE D'AOSTA	20	1
LIGURIA	18	18
LOMBARDIA	134	66
NORD EST	64	47
TRENTINO-ALTO ADIGE	36	3
VENETO	80	60
FRIULI-VENEZIA GIULIA	48	62
EMILIA-ROMAGNA	68	72
CENTRO	29	62
TOSCANA	27	55
UMBRIA	33	65
MARCHE	29	74
LAZIO	46	53
CAMPANIA	37	49
ABRUZZI-MOLISE	29	54
ABRUZZI	31	46
MOLISE	24	72
SUD	19	47
PUGLIA	14	46
BASILICATA	21	57
CALABRIA	30	40
SICILIA	19	54
SARDEGNA	46	19

Cont.

EC Member States and Regions	Nitrogen from cattle, sheep and pigs (kg N/ha. uaa)	Arable crops as percent. of uaa
LUXEMBOURG (GRAND-DUCHE)	113	43
NEDERLAND	231	42
NOORD-NEDERLAND	152	41
GRONINGEN	94	67
FRIESLAND	208	12
DRENTHE	135	53
OOST-NEDERLAND	296	34
OVERIJSSEL	278	32
GELDERLAND	310	35
WEST-NEDERLAND	141	48
UTRECHT	307	6
NOORD-HOLLAND	134	43
ZUID-HOLLAND	156	42
ZEELAND	38	85
ZUID-NEDERLAND	369	47
NOORD-BRABANT	403	43
LIMBURG	285	56
PORTUGAL	NA	64
UNITED KINGDOM*	63	38
NORTH	77	24
YORKSHIRE AND HUMBERSIDE	68	53
EAST MIDLANDS	49	71
EAST ANGLIA	33	87
SOUTH EAST	52	68
SOUTH WEST	97	41
WEST MIDLANDS	95	52
NORTH WEST	116	30
WALES	100	16
SCOTLAND	35	19
NORTHERN IRELAND	101	30
Maximum N for the listed regions	403	96

* UAA for regions of the UK is based on 1983 data.
Source: Eurostat (1988) Regions - Statistical Yearbook.

Figure 3.1 shows the intensity of nitrogen produced by cattle, sheep and pigs expressed in cow equivalents per ha. of utilised agricultural area. The figure complements the data presented in Table 3.3 and indicates clearly that at the "first level" regions in the Community, the highest rates of intensity are found in Belgium and the Netherlands. However, this can be misleading in that wide variation in this measure of intensity can occur at the sub-regional level.

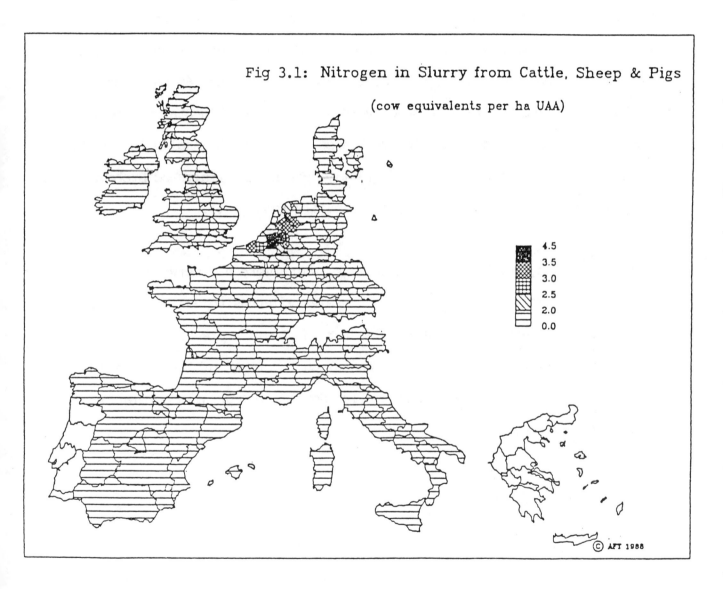

Fig 3.1: Nitrogen in Slurry from Cattle, Sheep & Pigs (cow equivalents per ha UAA)

3.2 Arable crops

Another source of nitrogen pressure is arable cropping, both because of the associated levels of chemical N inputs and because cultivation leads to mineralisation of N in the soil organic matter. The relative intensity of arable cropping per hectare of UAA is given in Table 3.3.

There is a higher intensity of arable cropping in Denmark than in other Member States, with 92 per cent of utilised agricultural area devoted to arable cropping. The next highest State values are 64 and 60 per cent for Portugal and West Germany respectively. Regions within States, however, have higher values, such as Ile de France (95%), East Anglia in the UK (87%), Zeeland in the Netherlands (85%), Marche in Italy and Brabant in Belgium (74%), Berlin West (71%), and Castilla-Leon (70%).

Certain regions, such as those in Denmark, which were not very intensive in terms of animal manure production are very intensive as regards arable cropping. Problems of nitrates in water in Denmark are due in a large measure to the high incidence of annual cultivation of the soil and are compounded by the small area nationally devoted to grassland relative to the livestock population.

Chapter 4

FERTILIZER USE PRACTICES IN THE EC

4.1. Evolution of nitrogen use in European Community

4.1.1. Fertilizer nitrogen

There has been a dramatic increase in fertilizer use in the EC particularly between 1950 and 1980. The use of P and K fertilizer increased rapidly after 1945 but has stabilised over the past 20 years. In contrast, N fertilizer use has continued to increase up to the present. The trend in N fertilizer use in Member States in recent years is summarised in Figure 4.1. This shows that Member States fall into three broad groups in relation to N consumption, (a) Netherlands with an average rate of over 200 kg N/ha; (b) Denmark, Belgium-Luxembourg and Federal Republic of Germany using between 100 and 150 kg/ha; and (c) other countries using between 25 and 90 kg/ha.

It is important to note that the values used for Figure 4.1 are based on average values from the individual Member States and that some farms and regions receive much higher levels of N while other farms and regions received little or no N fertilizer. The increase in nitrogen fertilizer use has been due to farmers adopting more intensive methods of farming, this has been helped by the tendency of agricultural advisory services to increase N fertilizer recommendations. Table 4.1 shows the trend in maximum N fertilizer recommendations by the advisory service for grassland in selected Member States.

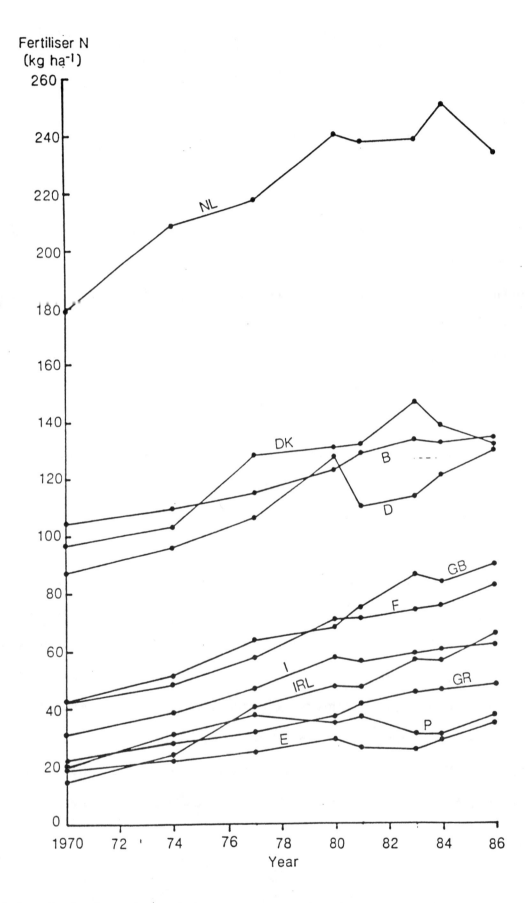

Figure 4.1. Evolution of fertiliser N consumption in European Community (Lee, 1987).

Table 4.1: Maximum fertilizer N recommendations by advisory services in selected Member States for intensive grassland (kg N/ha./pa)

	1960	1970	1980	1988
		Pasture		
DK	–	–	250	250*
D	200	300	350	380
IRL	185	225	375	390
NL	–	–	400	400
UK	190	–	300	275
		Silage		
DK	–	200	350	350
D	120	240	240	300
IRL	180	315	333	325
NL	–	–	400	400
UK	245	–	320	330

* DK recommendations for pasture are 100 kg lower than for silage because of the N in excreta of grazing animals.

Table 4.1 indicates that the fertilizer recommendation for intensive grassland has approximately doubled between 1960 and the present. Corresponding recommendations for cereals are summarised in Table 4.2.

Table 4.2: Maximum fertilizer N recommendations by advisory services in selected Member States for intensive cereal production (kg N/ha)

	1960	1970	1980	1988
		Spring Barley		
DK	75	90	120	130
F	70	100	140	177
D	70	120	160	170
IRL	50	85	125	140
NL	–	–	–	–
UK	80	–	125	125
		Winter Wheat		
DK	90	140	140	180
F	–	–	–	–
D	100	150	200	210
IRL	–	–	–	210
NL	–	–	–	200
UK	100	–	125	200

The values in Table 4.2 also indicate an approximate doubling of the N fertilizer recommendation for cereals between 1960 and the present. It can be noted that maximum recommendation for grassland is approximately double that for cereals.

The increase in fertilizer use shown in Figure 4.1. and the trend towards increased recommendations shown in Tables 4.1 and 4.2 are due to economic conditions where greater fertilizer use has been justified by the economic incentives and technical developments enabling farmers to increase production. In general, the value of the produce sold off the farm, such as meat, milk and cereals, has increased faster than the cost of fertilizer nitrogen. The guaranteed prices under CAP have been an important factor in maintaining prices and increased production.

From Figure 4.1 it can be seen that, on average, the original six Member States of the Community have the highest N consumption whereas the three most recent members of the Community (El,Es,P) have the lowest average consumption of N.

4.1.2. Manure nitrogen

An estimate of the trend in average nitrogen production in animal manures in kg per ha of UAA is summarised in Figure 4.2. The N production in animal manures is generally of the same order as N in chemical fertilizers in each Member State. A comparison of Figures 4.1 and 4.2 shows that countries with high N fertilizer use also have high N production in animal manures. Figure 4.2 also indicates that N in animal manure production has not increased as rapidly in recent years as fertilizer N use.

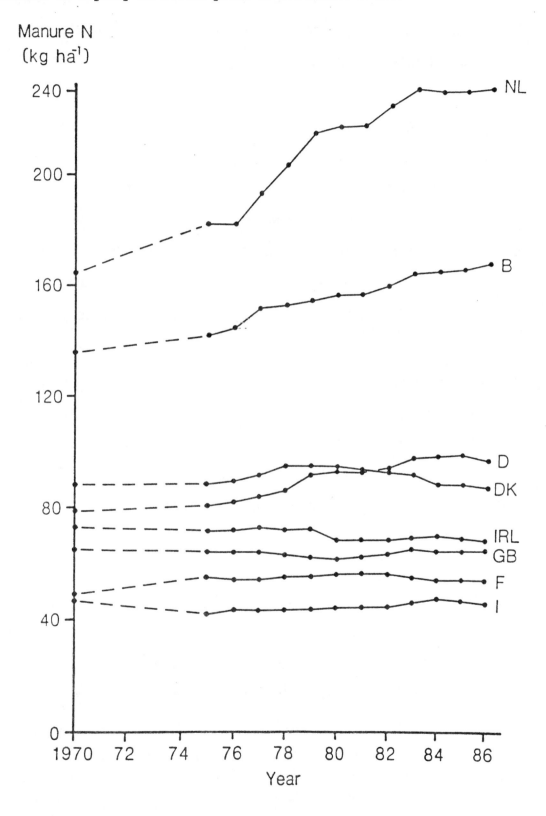

Figure 4.2. Production of N in cattle, sheep and pig manure.

It is possible that the increased N production in animal manures may be underestimated in Figure 4.2, as it is based on manure production with constant N content based on 1978 manure composition values. However, it is likely that the nitrogen content of manure may have been a little lower in 1960, for example, compared with the present.

There is a long term trend for an increase of fertilizer N relative to animal manure N being applied to agricultural land. This changing pattern over the past century is illustrated for Germany in Figure 4.3. It shows that the total N use has increased more than tenfold over this period and that the proportion in chemical fertilizer has increased from less than 10 per cent of total at the start of the century to about 50 per cent at present. A similar trend has taken place in the other Member States.

4.1.3. Future evolution of N use

It is not possible to predict accurately future nitrogen fertilizer use in the Community as this depends primarily on the price farmers receive for the animals and crops produced by N fertilizer and the price of the fertilizer N itself.

Figure 4.4 compares the trends in fertilizer N consumption in (a) Ireland -one of lowest in the EC, (b) the Netherlands -the highest- and (c) New Zealand. It shows that New Zealand agriculture, though well developed, uses very little fertilizer N in comparison to agriculture in the EC.

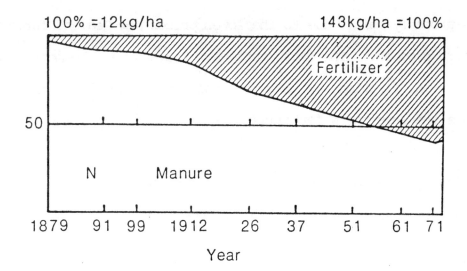

Figure 4.3. Changing N application in fertiliser and manure in Germany, over the past century (Flaig et al. 1978).

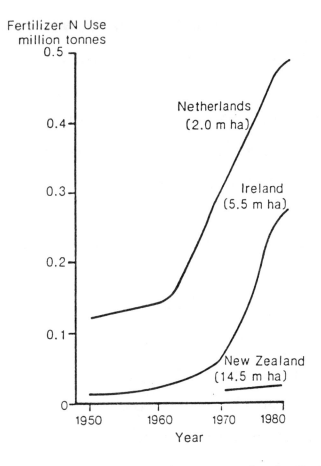

Figure 4.4 Comparison of trends in fertiliser N use in the Netherlands, Ireland and New Zealand (Tunney, 1985).

This is probably mainly due to the lower prices New Zealand farmers receive for their produce. The fact that New Zealand may be more climatically suited to N fixation by legumes than some parts of the EC may also influence the level of N fertilizer use.

In previous generations in Europe when N fertilizers were scarce and relatively expensive, farmers used legumes and animal manures as a source of nitrogen to maintain soil fertility and crop yields. At present, legumes are not generally used in intensive agriculture in the EC and animal manures are more often treated simply as a disposal problem than as an important source of plant nutrients.

If all farms in the EC were to intensify to the level of the most intensive farms, the total N use would be several fold what it is at present.

There is already some evidence in the current year (1988) that quotas on production and limits on agricultural spending are having an impact on stabilising or reducing N fertilizer use.

4.2 N fertilizer use and effect on crop yield and nitrate in water

There are many scientific publications showing the effect of N fertilizer on yield increase. Undoubtedly, N fertilizer has contributed greatly to increased food production, particularly on intensive farms in recent years. The relationship between average wheat yield and average N use in France over 30 years is illustrated in Figure 4.5.

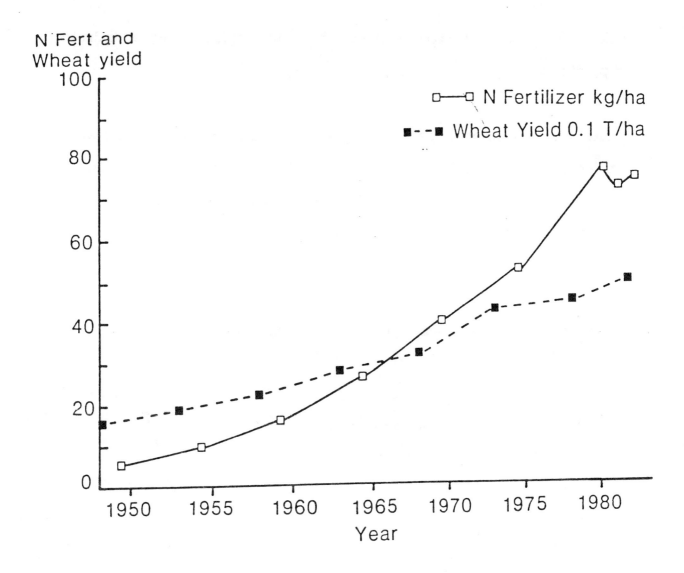

Figure 4.5. Average fertiliser N rate and wheat yield in France (Remy, 1985).

The average yield of wheat in France has increased from 1.5 t/ha in 1950 to 6.3 t/ha for 1984 while at the same time the average N fertilizer use increased from less than 10 to about 80 kg N/ha.

There is much discussion on how increasing N fertilizer rates impact on the nitrate content in water. There are some conflicting results; however, the general consensus is that increased N fertilizer use leads to an increased level of nitrates in water though the increase in nitrate relative to N use will be influenced by many factors including soil, climate, hydrology, crop, etc.

Figure 4.6 illustrates, as an example, the evolution of nitrate levels in surface water in small agricultural catchments in the Seine Basin in Normandy, France. It indicates an increasing level of nitrate in water reflecting the increased use of N fertilizer.

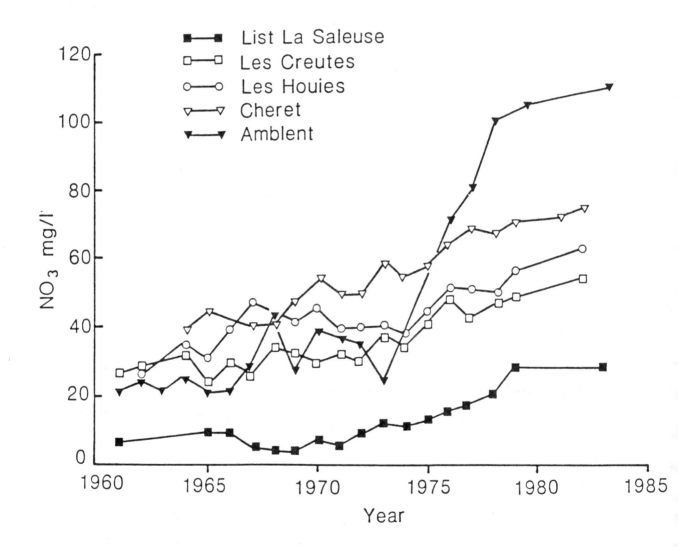

Figure 4.6. Evolution of water nitrate levels in a small agricultural catchment of the Seine basin (Remy, 1985).

4.3 Recent developments in nitrogen fertilizer use in the EC

This section is based on submissions of EC Member States to the "European Study Days in Water Pollution and Intensive Farming", CEPFAR, Brussels, March 23-25, 1988. An extensive summary of each Member State's submission relating to this section is contained in APPENDIX 4 of the report. Opinions or comments contained in the appendix summaries reflect those of the Member State delegations concerned and do not necessarily represent the views of the study team.

4.3.1 Chemical Fertilizers

EC Member States are becoming increasingly aware of the role of chemical nitrogen (N) fertilizers in increasing nitrate levels in drinking water. This is gradually leading to a more rational use of N fertilizers based on experimental results for the varying soil and climatic zones within each area. However, excessive rates of N are still being applied in some regions.

Most Member States now recommend that N should not be used in the autumn on winter cereals. More account is being taken of the role of crop rotations in determining the amount of soil nitrogen released and its importance in optimising the use of chemical N fertilizers. In Denmark, the target for N fertilizer application is the economically optimal quantity of N per field, and to achieve this goal, all practical management methods, including N-forecasts based on temperature and precipitation data for the September-March period, soil N-min analysis and plant analysis are used. In addition, all farmers must be able to produce a fertilization plan at the request of the authorities.

In grass and maize production, not enough account is being taken of the nitrogen in animal manures used when chemical N fertilizers are being applied to these crops. This is leading to an excessive use of N in some locations, e.g. in parts of the Netherlands.

4.3.2 Animal Manure

In some EC regions, excessive livestock intensification has led to over-production of effluent in relation to the amount of land available for spreading it. This is contributing to increased nitrate levels in drinking water. The problem becomes more acute when there is a long indoor period, inadequate manure storage, poorly drained or pervious soils and large imports of animal feed independent of farm size. Farmers in these so-called "manure surplus" regions must find land resources for manure spreading outside their farms and sometimes rather far away from them. Animal manure is applied principally to grassland and fodder maize crops.

REFERENCES for Chapter 4

Flaig, W., Nagar, B., Sochting, J. and Tietjen, C. (1978) Organic Materials and Soil Productivity, Soils Bulletin No. 35, FAO, Rome.

Lee, J. (1987) "Land Resources and their Use in the European Community", Scientific Basis for Soil Protection in the European Community, (Ed.) H. Barth and P.L'Hermite, Elsevier Applied Science Publishers, London.

Remy, J.C. (1985) "Quleques Elements pour une Amelioration de la Gestion des Fertliisants", Environment and Chemicals in Agriculture, (Ed.) Winteringham, F.P.W., Elsevier Applied Science Publishers, London, pp 163-206.

Chapter 5

THE NITROGEN CYCLE AND POLLUTION

The nitrogen cycle represents a complicated series of interactions between man, animals, plants, soils, air and water. For the purposes of this study, the main factors to be considered are the inputs into the soil/plant system and in particular the effects of altered inputs on crop yield and on nitrogen losses through leaching of nitrate to groundwater. A simplified diagram of pertinent parts of the nitrogen cycle is shown in Figure 5.1.

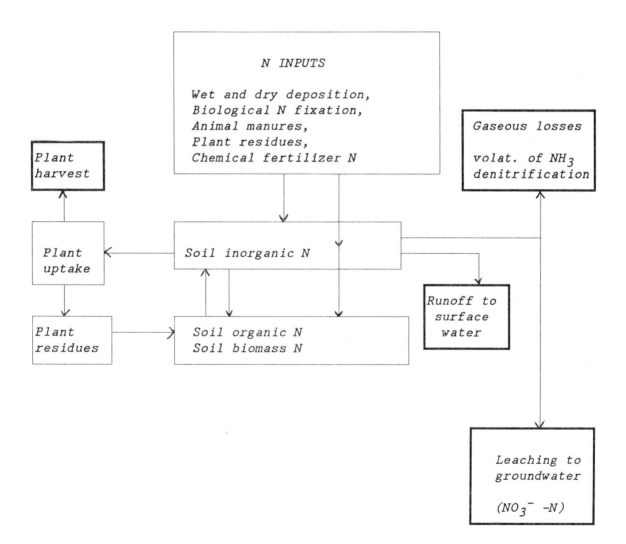

Figure 5.1: Simplified nitrogen cycle

5.1 Nitrogen inputs

Wet and dry deposition:

NH_3 and NO_x gases in the atmosphere may be precipitated in rainfall and dust particles, in amounts varying from 10 to 50 kg N/ha depending on the presence or absence of intensive livestock production units in the area and on emanations from industrial units and automobiles.

Biological N fixation:

N fixation by soil micro-organisms can vary from 10 to 20 kg N/ha while N fixation by legumes may be as high as 300 kg N/ha. However, biological N fixation decreases as inorganic N supply increases and would be expected to be low at the upper limits of the added N allowed under the proposed EC Directive.

Animal manures:

The amount of N applied in farm wastes varies greatly depending on the intensity of animal production and cropping practice. Excessive rates are generally applied where insufficient land is available for spreading pig and poultry manure. Animal manure N consists of about 50 percent organic N and 50 per cent ammonium - N. A considerable amount of the NH_4^+ -N may be lost through volatilisation of ammonia especially where manures are not incorporated into soil after spreading.

Plant residues:

N contribution from plant residues depends on agricultural practice and/or the carbon : nitrogen ratio of the residues. As much as 200 kg N/ha can be released through mineralisation when permanent grassland is ploughed, whereas the addition of high carbon residues such as straw, may result in temporary immobilisation of fertilizer or soil N into soil biomass.

Fertilizer N:

N application rates vary with agricultural practice, cropping regime, stage of rotation etc. Forms of N applied are urea, ammonium and nitrate N.

Soil N:

Soil N is an indirect input which arises from mineralisation of soil organic N by soil micro-organisms. It may vary from 30 to 150 kg N/ha depending on soil type and previous cropping history. It is not always possible to distinguish between soil N and slow release of N from previous applications of animal manure or plant residues.

5.2 Sources of nitrogen losses and export

Surface run-off:

Up to 50 per cent of added fertilizer N or animal manure N may be lost in surface run-off if rain causing run-off occurs within the first 48 hours after spreading on wet soils. This source of loss can be largely eliminated by avoiding spreading on wet or frozen soils and by checking weather forecasts to ensure that heavy rain is not imminent.

There may also be loss of nitrate to surface water through subsurface run-off but it is less well understood and is difficult to quantify.

Gaseous losses:

The two main pathways of N loss in this respect are, (1) Volatilisation of ammonia following surface spreading of animal manures without subsequent incorporation into soil and also following spreading of urea on freshly limed soils, open pastures or bare soils under drying conditions.

(2) Denitrification (i.e. $NO_3^- \longrightarrow N_2O \longrightarrow N_2$) by soil bacteria occurs under anaerobic or waterlogged conditions provided there is a carbon energy source available. Soil type, including clay content, has a considerable influence on denitrification.

Informed management can help reduce gaseous losses.

Plant harvest:

N uptake and yield response by the plant is principally dependent on temperature, radiation and water supply. It is important to match fertilizer applications to plant growth conditions and potential.

The efficiency of recovery of fertilizer N in harvested plant material is generally 50-60 per cent in cereal crops and 50-80 per cent in cut grassland.

Leaching of nitrate to groundwater:

Leaching of nitrate takes place mainly in autumn/early winter. In cereal crops, N uptake ceases in July but N from soil organic matter and plant debris continues to be mineralised and nitrified due to favourable conditions. Nitrate accumulates in the topsoil while the soil is in moisture deficit but as soon as rainfall exceeds evapotranspiration, the accumulated nitrate leaches through the soil profile. In clay soils, considerable amounts of nitrate - N may be denitrified.

Leaching of nitrate from grazed grassland is mainly from dung/urine patches delivered in September, October and November. Animal manures applied to either grassland or tillage soils in autumn/early winter will lead to leaching of nitrate.

The amount of nitrate which leaches is a function of the crop, soil and fertilizer practice. The concentration of N in leaching water is a function of the volume of leaching water i.e. rainfall minus evapotranspiration. The time interval before leaching nitrate reaches groundwater is a function of the volume of leaching water, soil type and the depth of soil over-burden, which may range from less than one metre to greater than 60 metres.

Chapter 6

MODEL FOR NITROGEN USE

6.1 Chemical fertilizer equivalent of animal manures

We wish to estimate the effects of altering the rate of nitrogenous fertilizer use on (a) income from farming and (b) nitrate concentration in groundwater. The nitrogenous fertilizer may come from chemical fertilizer or from animal manures. Animal manure applications would need to be converted to the 'equivalent chemical fertilizer nitrogen application rates', so that nitrogen from both sources could be treated as additive. The conversion from animal manure nitrogen to chemical nitrogen equivalent would have to take account of manuring practices. These practices relate to animal origin of manures, manure storage, application rates to crops and the seasonal timing of manure applications. These practices will influence the nitrogen content of manures, the response in crop growth and nitrogen uptake, as well as the possible contribution to water pollution. While there is a wide variation possible in practices, analysis could be based on typical or recommended practices. Such simplification could be justified as, in areas experiencing nitrate pollution problems, there is likely to be greater care in the storage and spreading of animal manures.

We need a set of relationships which give the contribution of animal manures, expressed as chemical nitrogen equivalent (NE), in kg per hectare. In general these relationships can be represented as follows:

$$NE = f(\text{animal source}, \text{manuring practice})$$

6.2 Efficiency of nitrogen in animal manures

The efficiency of nitrogen in animal manure is less than that of chemical fertilizers[1]. Part of the nitrogen, in the form of ammonium, can readily volatilise. Part of the nitrogen is fixed in the formation of humus, although eventually, under a given manuring practice, soil attains an equilibrium humus content. (At equilibrium the amount of nitrogen fixed in the humus equals that liberated from the humus by mineralisation). The nitrogen in the organic matter is liberated only after mineralisation and much of this occurs in late summer and autumn after crops have finished taking up nitrogen.

Crops which have nitrogen uptake patterns that better match the release of nitrogen in the soil will give rise to lower nitrate leaching. Soil cultivation is a major factor influencing the release of nitrogen from the soil's reserve of organic matter. Cultivation stimulates breakdown of the organic matter and release of its nitrogen. Nitrogen lost through leaching consists primarily of soil nitrogen mineralised outside the growing season. The extent of this leaching also depends on the length of the growing season, which increases for the following crops in the order: cereals, potatoes, beets and grass. With arable crops it is helpful if crops which take up nitrogen over the autumn/winter period, such as winter cereals or catch crops are planted after the main crop is harvested.

---

[1] Commission of the European Communities (1978a) The Spreading of Animal Excrement on Utilized Agricultural Areas of the Community, Volume I, Information on Agriculture, No. 47.

Kolenbrander[1] reports on a "nitrogen efficiency index" for animal manures, which expresses the yield response to nitrogen from animal manures relative to that from chemical fertilizer nitrogen. The average values found in field experiments are given in Table 6.1 and are in agreement with annual efficiency indices for grassland given by Schechtner et al[2]. The nitrogen efficiency index is lower for grassland than for arable land "due to higher losses by volatilisation of ammonia on grassland, whereas about 50 per cent of the difference between spring and autumn application is caused by leaching"[3].

Table 6.1: Yield response from animal manure nitrogen relative to yield response from chemical fertilizer nitrogen

Animal Manure (% Mineral N)	Arable land		Grassland*	
	Spring	Autumn	Spring	Autumn
Liquid manure (94%)	80%	40%	70%	35%
Slurry (50%)	50%	25%	35%	20%
Farmyard manure (10%)	40%	20%	20%	10%

* Zero grazed
Source: Kolenbrander (1981) op. cit. Table 4.

Table 6.1 reveals a higher efficiency index for manures with higher mineral

---

[1] Kolenbrander, G. J. (1981) "Leaching of Nitrogen in Agriculture", Nitrogen Losses and Surface Run-off from Landspreading of Manures, (Ed.) Brogan J.C., Martinus Nijhoff/Dr. W. Junk, The Hague.

[2] Schechtner, G., Buchgraber, K. and Eder, G. (1988) "Economical Slurry Application on Grassland", Paper presented at the Joint FAO-subnetwork 4 and EC-cost Workshop, held at FAC Liebefeld, Bern, Switzerland, 18-22 June.

[3] Kolenbrander (1981) op. cit. p 213.

nitrogen content. If efficiency indices could be based on the mix of mineral/organic nitrogen in manures, this would facilitate estimating efficiency indices for slurry from different animals. Such a basis of estimation was set out a report to the Commission in 1978[1]. Estimates using this basis (see Appendix Tables 6.1 and 6.2) would indicate that the values in Table 6.1 relate to response to animal manure nitrogen in the year after application. However, conformity with Kolenbrander's results in Table 6.1 would require a higher contribution from mineral nitrogen and less from organic nitrogen, (see Appendix Table 6.2)

Conformity with Kolenbrander's results can be achieved for arable crops by increasing the coefficient for mineral nitrogen by 10 per cent (equivalent to reducing volatilisation on application from 20% to 12%) and reducing the coefficient for organic nitrogen by 50 per cent (equivalent to assuming that 25% rather than 50% is mineralised in the first year). In the case of grassland, the coefficient for mineral nitrogen will also be increased by 10 per cent (equivalent to reducing volatilisation after application from 32% to 25%), while the coefficient for organic nitrogen will be reduced by 75 per cent for autumn application and set to zero for spring application. Data in Table 6.1 relate to zero grazing, but under grazing conditions manure is excreted directly on the pasture and volatilisation losses are higher, 42 per cent as opposed to 12 per cent. Applying this modified basis gives the estimated efficiency indices as outlined in Table 6.2.

---

[1] Commission of the European Communities (1978a) op. cit. pp 16 and 26.

Table 6.2: Estimates of the efficiency index (%) for nitrogen from animal slurry*

Slurry type (% mineral N)		Arable land		Grassland		
				Zero grazed		Grazed
		Spring (%)	Autumn (%)	Spring (%)	Autumn (%)	(%)
Cattle slurry	(40%)	44	22	30	18	20
Pig slurry	(50%)	52	25	38	20	25
Poultry slurry	(70%)	66	29	53	24	35
Calf slurry	(80%)	73	31	60	26	40

* Based on modifications to coefficients in CEC Report 1978a op. cit. (pp 16 and 26), to make efficiency indices conform to the estimates in Table 6.1.

A systematic basis for estimating efficiency indices is a desirable approach, because it can be used for manures in different forms and from different animals. In principle, the concept of chemical nitrogen equivalent of animal manures is operational for crop yield response. Nitrogen efficiency indices can be used to convert nitrogen in animal manures into its equivalent in chemical nitrogen (NE). Thus, for example, 100 kg of cattle slurry nitrogen applied to zero grazed grassland in spring would be equivalent to 30 kg of chemical nitrogen, based on a nitrogen efficiency index of 30 per cent (see Table 6.2).

The estimates of nitrogen efficiency for crop yield in Table 6.2 are estimates of a general level of efficiency, but the actual efficiency in practice is very variable. Differences in volatilisation of nitrogen is a major source of variation in nitrogen efficiency. Volatilised nitrogen will be dispersed in the atmosphere and will not give rise to concentrations of nitrate in the groundwater. However, atmospheric pollution is not desirable and more efficient use of nitrogen from animal manures is desirable so as to minimise the need to add chemical nitrogen in order to achieve optimum crop yields. The level of efficiency can be increased by application methods which reduce volatilisation. Dilution of slurry gives a dramatic increase in nitrogen efficiency (e.g. from 35% to 60% for cattle slurry and 50% to 70% for pig slurry when diluted 1 : 1[1]). Kiely[2] indicates that band spreading of slurry on grassland can increase efficiency up to 80 per cent by reducing volatilisation. Vetter et al[3] suggest that "under optimal application conditions .... nitrogen in slurry can be up to 80 - 90 per cent as effective as nitrogen in mineral fertilizers".

It is clear, therefore, that efficiency indices need to be established for specific manuring practices. Furthermore, efforts to reduce pollution and increase the effectiveness of animal manures, particularly in areas with water pollution problems, will lead to farming practices that give higher efficiencies than those in Table 6.2. The manurial practices will reflect the livestock mix and density, as well as the cropping pattern found on farms. Hence, it is to be expected that chemical nitrogen equivalent (NE)

---

[1] Schechtner et al (1988) op. cit. Table 2.

[2] Kiely, P.V. (1988) "Effect of Spreading Method on Slurry Nitrogen Utilisation by Grassland", *Proceedings of 12th General Meeting of the European Grassland Federation*, Dublin. pp 353-357.

[3] Vetter, Heinz, Steffens and Gunter (1988) "*Guidelines for an Economic Use of Slurry on Agricultural Land*", Paper presented at the Joint FAO-subnetwork 4 and EC-cost and EC-COST Workshop, held at FAC Liebefeld, Bern, Switzerland, 18-22 June.

of animal manures will differ between farms and between regions. When animal manuring practices are represented as equivalent rates of chemical nitrogen, total nitrogen application for crop yield can be represented as:

N = NE + SN,

    where N is total nitrogen in kg/ha,
        NE is the chemical nitrogen equivalent of animal manure
  and SN is chemical nitrogen in kg/ha.

6.3 Crop response to Nitrogen

We need to estimate crop yield responses to total nitrogen application, in order to estimate the effects on crop yield and income of altering the application rate. While nitrogen may be applied at different times of the growing season, we will use an annual rate of application, assuming typical or recommended seasonal timing of applications. Hence, the yield/ nitrogen relationship will be based on a one year period.

For the purpose of illustration assume the commonly used quadratic relationship as follows:

$$Y = a + b*N - c*N^2 \quad \dots\dots\dots\dots\dots\dots\dots\dots\dots\dots\dots\dots\dots\dots\dots\dots\dots\dots\dots(1).$$

where Y is yield in dt/ha
and a, b and c are technical coefficients.

Note that * where used in equations, indicates multiplication
This quadratic function has yield increasing at a decreasing rate as the rate of nitrogen application increases. The positive yield at zero nitrogen implies that uptake of nitrogen by the crop exceeds nitrogen

applied up to a certain level of application, which we will call the balance level of nitrogen application (NB). The balance level of nitrogen application is the level where nitrogen applied equals nitrogen uptake by the crop. At the balance level, nitrogen applied does not contribute to nitrogen available for water pollution. As nitrogen application increases linearly, crop yield, and hence nitrogen uptake, increase at a declining rate. As nitrogen application is increased beyond the balance rate, an increasing proportion of the nitrogen applied is available to contribute to water pollution. While the nitrogen content of the crop can increase somewhat at higher levels of application, we will use the approximation of a constant nitrogen content for the crop produced. The nitrogen content of the crop can be represented by:

$$NC = k*Y \quad \dots\dots\dots\dots\dots\dots\dots\dots\dots\dots\dots\dots\dots\dots\dots\dots\dots\dots(2)$$

where NC is the nitrogen in the crop yield (kg per ha), and
k is the nitrogen content of the crop (kg per 100 kg).

The balance rate of nitrogen application (NB), is the level that satisfies the condition:

$$\begin{aligned} N &= NC \\ &= k*Y \\ &= k(a + b*N - c*N^2), \text{ for the quadratic model} \\ \text{or } 0 &= k*a + (k*b - 1)*N - k*c*N^2 \end{aligned}$$

Solving this equation for N would give the balance rate (NB) of total nitrogen application:

$$NB = \{(1 - k*b) + \text{or} - [(k*b - 1)^2 + 4*k^2*a*c]^{0.5}\}/2*k*c \dots\dots(3)$$

such that NB>0,
where NB is the rate of nitrogen application such that NC = N.

Agricultural production can contribute to nitrate pollution of groundwater only when the nitrogen application rate exceeds the balance rate, that is, when N > NB.

6.4 Contribution of nitrogen application to nitrates in groundwater

When nitrogen application exceeds the balance rate, the nitrogen not taken up by the crop is potentially available to contribute to nitrate pollution of water. The nitrogen not taken up by the crop is represented as:

$$NA = N - NC,$$

where NA is nitrogen not taken up in crop yield.

For the quadratic model (see equations 1 and 2) this is:

$$NA = N - k(a + b*N - c*N^2),$$
$$= -k*a - (k*b - 1)*N + k*c*N^2 \quad \text{...........................(4)}$$

The relationship between the concentration of nitrates in the drainage water and the applied nitrogen available for leaching (NA) needs to be established.

For given soil and climatic conditions a linear relationship is likely and might be represented by:

$$NO3 = S + w*NA,$$

where   NO3 is nitrate leaching (mg/litre of drainage water),
        S is NO_3 from sources other than nitrogen applied, and
        w is a technical coefficient.

Substituting the formula for NA from equation 4 into this equation gives:

$$\begin{aligned}
NO3 &= S + w*[N - k*(a + b*N - c*N^2)] \\
&= (S - w*k*a) + (w - w*k*b)*N + w*k*c*N^2 \\
&= d + e*N + f*N^2, \quad \dots\dots\dots\dots\dots\dots\dots\dots\dots\dots\dots\dots\dots(5)
\end{aligned}$$

where the technical coefficients are $d = S - w*k*a$,
$e = w - w*k*b$,
$f = w*k*c$.

6.5 Nitrate leaching from animal manures

The mineral nitrogen component in animal manures contributes to both plant growth and leaching, in the same way as chemical nitrogen fertilizers, but that is not the case for the organic nitrogen component. Organic nitrogen in animal manure is only released following mineralisation. With spring applications, for example, much of the mineralisation occurs late in the growing season and consequently the uptake by plants of nitrogen from the organic source is less than for mineral nitrogen. The uptake of nitrogen released from organic matter applied in the spring varies between crops, being about 95 per cent for permanent grassland, but less for arable crops, ranging from 50 per cent for cereals to 75 per cent for beets, (CEC 1978a, p 27). The lower uptake of mineralised nitrogen was taken into account in estimating the nitrogen efficiency indices for converting animal slurry nitrogen to its chemical equivalent. In contrast, the contribution of organic nitrogen to leaching is higher than for mineral nitrogen applied in spring. This is due to the relatively high nitrogen release in the autumn of the year and the high rainfall in autumn and winter. Hence, nitrogen mineralized from the organic matter in animal manures can contribute to nitrate leaching to a greater extent than would be indicated by its nitrogen efficiency index for crop yield. Therefore, the full impact of animal manures on nitrate leaching may not be captured by their impact on chemical nitrogen equivalent (NE). Where mineralisation of organic nitrogen applied as animal manure leads to higher leaching, the relationship (equation 5) between nitrate concentration in drainage water (NO3) and total nitrogen application (N), would need to be modified. The modified relationship can be represented as follows:

$$NO3 = d + e*N + f*N^2 + g*(AN) \dots\dots\dots\dots\dots\dots\dots\dots\dots(6).$$

where AN is the total nitrogen per hectare from animal slurry, and
g is a technical coefficient appropriate to the animal slurry

The size of the coefficient g depends not only on the amount of drainage water but also on the organic nitrogen content of the manure and the extent to which it is mineralized in the first year after application. Table 6.3 gives estimates of the percentage of organic nitrogen applied as animal manures which is lost as additional leaching and is represented by g*(AN) in equation 6.

Table 6.3: Estimates of the percentage of organic nitrogen applied as animal manures which are lost as additional leaching (represented by g*(AN) in equation 6)*

Time of application	Arable land (%)	Grassland (zero grazed) (%)
Spring	12	0
Autumn	7	1

* These estimates are based on Report to CEC (1978a) op. cit. (pp 16 and 26) modified according to the nitrogen efficiency index estimates in Table 6.2.

Because of a lower rate of mineralisation and better crop cover in autumn and winter, animal manures applied to grassland would not be expected to lead to any significant additional nitrate leaching due to the organic nitrogen content of animal manures. This implies that for grassland the technical coefficient g in equation 6 would be expected to be zero. Under arable cropping the nitrogen from the organic matter in animal manures contributes to nitrate leaching to a greater extent than would be indicated by its nitrogen efficiency index for crop yield. Hence the coefficient g in equation 6 would be expected to be positive for arable crops.

Arable cropping, because of greater mineralisation of organic nitrogen, is characterised not only by higher nitrate leaching at zero nitrogen application, but also by higher leaching from the organic nitrogen of applied animal manures. The nitrogen efficiency index of animal slurry is also higher for arable crops than for grassland (Table 6.2). These factors contribute to higher leaching of nitrogen from arable crops. The leaching of mineralized nitrogen from arable cropping can be reduced by crop rotations which provide crop cover in the autumn and winter. However. lowest leaching losses would arise if animal manures were applied to permanent grassland in the March to July period. It should be noted, however, that low mineralisation of manure organic matter in the first year means a build up of soil organic matter. If this were released by mineralisation in subsequent years it would result in higher leaching, and crop yield, at zero nitrogen application. Subsequent cultivation of soil with a high organic matter content would give rise to rapid release of this nitrogen reserve. Hence, the permanence of grassland is important in relation to nitrate leaching.

6.6 Persistence of leached nitrates in the groundwater

The nitrates leached into the drainage water may not all persist as nitrates in the groundwater, so that an adjustment may be required. Nitrates that persist (NO3P) so as to reach the groundwater are represented as follows:

$$NO3P = p*NO3,$$

where NO3P is nitrates that persist and reach the groundwater, and
p is the fraction of leached nitrates that persists in the groundwater.

Substituting the formula for NO3 from equation 5 (equation 6 would be appropriate for arable crops receiving animal manures) into this equation gives:

$$NO3P/p = d + e*N + f*N^2 \quad \ldots\ldots\ldots\ldots\ldots\ldots\ldots\ldots\ldots\ldots\ldots\ldots(7)$$

Therefore, given a certain desired limit to nitrate concentration in groundwater (DNO3), the corresponding limit (NL) to be placed on nitrogen application is derived by solving the above relationship for N, when NO3P is set at the desired level, DNO3. The solution is given by the following relationships:

$$0 = (d - DNO3/p) + e*NL + f*(NL)^2,$$

$$NL = \{- e + or - [e^2 - 4*f*(d - DNO3)]^{0.5}\}/(2*f) \quad \ldots\ldots\ldots\ldots\ldots(8)$$

such that NL>0.

It should be noted that the value of NL depends not only on the desired level of nitrate concentration in the groundwater (DNO3), but also on nitrates coming from sources other than applied nitrogen, such as soil nitrogen released by mineralisation. Higher nitrates from these other sources would result in a higher value for the parameter d, (from equation 5, $d = S - w*k*a$)

The relationships above illustrate how crop yield and nitrate concentration in groundwater can be expressed as functions of the rate of nitrogen application. The quadratic example shows yield increasing at a decreasing rate, but nitrate concentration in the groundwater increasing at an increasing rate, as the nitrogen application rate is increased. The functional form of the yield model and its actual coefficients (a,b,c, and k in the quadratic example) will depend on the crop, on soil and climatic conditions. The extent to which nitrogen not taken up by the crop (NA) will contribute to high nitrate levels in drainage water (i.e. the coefficient w) will also depend on soil characteristics (especially porosity) and on climate (especially rainfall). Account must also be taken of the degree to which nitrates leached in the drainage water persist (i.e. coefficient p in equation 7) and are ultimately found in the groundwater.

6.7 Impact of Nitrogen Fertilizer Use on Farm Income

If nitrogen application is restricted to ensure water quality, this will affect crop yields and thereby output and incomes from farming. In an unrestricted situation, a farmer could choose the nitrogen application rate which would maximise income. In the case of cash crops, reducing the rate of nitrogen application would not alter other crop inputs, so the only cost saving would be in respect of the nitrogen input. In the case of crops produced for animal feed, additional crop yield is generally associated with additional costs besides nitrogen, related to additional animals. Hence, it is the margin over non-crop costs (i.e. costs excluding those costs associated with crop production) that has to be compared with the change in nitrogen cost, when assessing the optimum application of nitrogen. The unrestricted optimum yield is at that nitrogen application rate for which the value of additional yield is just offset by the cost of the extra nitrogen which induced the additional yield.

This condition is given by:

$$V*(dY/dN) = CN$$

where V is the value per unit of additional yield produced, and CN is the cost per unit of chemical nitrogen.

Returning to the quadratic example, this condition is:

$$V*(b - 2*c*N) = CN$$

which when solved for N gives the optimum level of total N (ON) as:

$$ON = (b - CN/V)/(2*c) \quad \dots\dots\dots(9)$$

where ON is the optimum level of N.

Equation 9 shows that the optimum rate of nitrogen application (ON) depends on fixed technical parameters (b and c of equation 1) of the production function and the ratio of nitrogen cost to product value (CN/V). Doubling the price of nitrogen would double the price ratio (CN/V), as would halving the product value and hence both would have the same impact on optimum nitrogen use and crop yield. As nitrogen costs are a small fraction (usually less then 10%) of the value of farm produce, doubling nitrogen price would have much less impact on profit and farm income than would arise from halving the product value. Hence, if price policy were to be used to reduce the use of nitrogen, while minimizing the impact on farm income, increasing nitrogen price is more effective than reducing product price.

It has been shown in equation 8 how the appropriate limit to nitrogen application (NL) can be derived, so as to ensure that a crop's contribution to nitrates does not bring nitrate concentration in groundwater above its desired level (DNO3). If the limit on nitrogen use was set at or above the optimum rate (ON), then it could not lead to any loss in potential income. If the purpose of restricting nitrogen use is to avoid contributing to nitrate pollution of water, then only rates of application which exceed the balance level (NB) are of concern. Hence, relevant restrictions on nitrogen application rate will lie in the range NB to ON.

When the nitrogen limit lies in this range (NB < NL < ON), the potential loss in farm income per hectare of crop (PLFI), is given by:

$$\text{PLFI} = V*(\text{YON} - \text{YNL}) - \text{CN}*(\text{ON} - \text{NL}) \dotfill (10)$$

where YON is yield with optimum nitrogen application (N = ON),
YNL is yield with nitrogen at the limit rate (N = NL), and
PLFI is potential loss of farm income per hectare of crop.

The actual loss may be smaller, indeed is likely to be smaller, if many farmers are not using the optimum level of N. However, in terms of the options facing farmers, their potential for earning income would have been reduced by PLFI. Since restrictions on nitrogen use would impact on future practice, it is their effects on production possibilities which are relevant. It might be argued that people should be compensated for such reduction in potential, at least for a transitional period so as to facilitate adaptation to new administrative restrictions.

6.8 Multipliers for income from farming

The potential loss of farm income (PLFI) estimated above applies only to potential loss in value added at farm level. However, an initial "direct" impact on income from farming would be associated with additional impacts, which are differentiated into "indirect" and "induced"[1]. The "indirect" effects arise from "downstream" economic activity involved in transforming the associated farm output into final agricultural output. The complete impact throughout the economy would also include effects "induced" by the change in spending arising from the additional household income. This expenditure would induce further income and expenditure changes, until the

---

[1] Henry, E.H. (1986) Multisector modelling of the Irish economy, Paper No. 128, Economic and Social Research Institute, Dublin p 134.

multiplier process had worked itself out. The "direct" impact on income from farming has to be adjusted by a "Moore-type" multiplier, which will give an estimate which includes both "indirect" and "induced" effects.

Henry[1], using an input/output model, has estimated household income multipliers per unit of final output for the agricultural sector in the Republic of Ireland. These multiplier estimates are given in Table 6.4 below, along with the Moore-type multipliers derived from them. Moore-type multipliers are the factors by which "direct" income effects are multiplied so as to include "indirect" and "complete" income impacts.

Table 6.4: Household income multipliers, per unit final output, for the agricultural sector of the Republic of Ireland and derived Moore-type multipliers, various years

	1968	1978	1982
Output multipliers:			
1. Direct	.5861	.4998	.3819
2. Direct + indirect	.7240	.6426	.5236
3. Direct + indirect + induced	1.1420	.8392	.6478
Moore-type multipliers:			
4. Indirect multiplier (2/1)	1.2353	1.2857	1.3710
5. Complete multiplier (3/1)	1.9485	1.6791	1.6963

Source: Henry, (1986) op. cit. Table 5.15

The Moore-type multiplier estimates indicate an increase in the indirect effects, as would be expected to arise due to the relative increase in downstream activity. The complete multiplier has declined, however, which could be explained by increased import intensity of induced activities. Its magnitude is estimated at about 1.7 in recent years.

---

[1] ibid.

The multipliers given above are for the economy of the Republic of Ireland. The corresponding multipliers for rural areas would be smaller. Henry[1] states that "Regional multipliers are generally much smaller than national ones, because a region is generally far more import-intensive than a nation". Estimates of 1983 multipliers for household income for three of the more rural regions in Ireland are given in Table 6.5.

Table 6.5: Household income multipliers, per unit final output, for the agricultural sector of three regions in the Republic of Ireland and derived Moore-type multipliers, 1983

	West	Donegal	Sligo-Leitrim
Output multipliers:			
1. Direct	.6192	.3843	.4078
2. Direct + indirect	.6833	.4335	.4455
3. Direct + indirect + induced	.8280	.5059	.5211
Moore-type multipliers:			
4. Indirect multiplier (2/1)	1.1035	1.1280	1.0924
5. Complete multiplier (3/1)	1.3372	1.3164	1.2778

Source: Henry, (1987) private communication

The Moore-type multipliers estimated for the State in 1982 are higher than the corresponding estimates for regions in 1983. The complete multiplier for regions was 1.3 compared with 1.7 for the State.

The magnitude of the multiplier will vary between regions and States, depending in particular on the degree of processing of farm produce and the

---

[1] Henry, E.H. (1984) "Input-output Analysis with reference to Agriculture and the Food Industries: a Comment", Agricultural Economics Society of Ireland, Proceedings 1983/84, pp 106-131.

extent of purchase of farm inputs. In the Netherlands, for example, Harthoorn[1] estimated the agriculture-induced value added for 1981 to be 2.25 times the value added in agriculture, forestry and fishing. This is 0.4 higher than the 1.7 estimated for the Republic of Ireland. In general, we might expect complete multipliers for Member States to be in the range 1.7 to 2.3 and for rural regions to be in the range 1.3 to 1.8.

The potential loss in total income per hectare of crop (PLTI) is got by adjusting the potential loss in farm income by the complete multiplier, as follows:

$$PLTI = M*(PLFI) \dots\dots\dots\dots\dots\dots\dots\dots\dots\dots\dots (11)$$

where PLFI is the potential loss in farm income per hectare, and M is the complete Moore-type income multiplier.

6.9 Possible savings at community level

Limits on nitrogen use which reduce farm production would, in the case of surplus products, also reduce expenditure involved in storing or exporting surpluses. Such savings for the Community could be set against the losses due to lower production. Savings at EC level will arise only for products which are benefitting from market support. The savings in public expenditure, arising from reduced nitrogen use, will depend on the market support regime and the supply/demand balance, both within and outside the

---

[1] Harthoorn, R. (1986) <u>Backward and Forward Linkages with an Application to the Dutch Agro-chemical Complex</u>, Netherlands Central Bureau of Statistics, National Accounts Occasional Paper Nr. NA-011, Table 1.

Community. In the case of dairy farms, for example, reduced nitrogen use would lead to lower stocking density, but stock reductions would occur mainly in non-dairy grazing livestock, while milk quotas would continue to be filled. Hence savings in EC expenditure would not occur for milk but would arise for products from other grazing livestock. For non-quota products, such as cereals, the savings could be based on the export refunds which would be saved if production were lower. The saving on expenditure on exports (SXEX), per hectare of crop, can be got by applying the savings per unit of product (SXUP) to the decrease in yield, as shown in equation 12.

$$SXEX = SXUP*(YON - YNL) \quad \ldots\ldots\ldots\ldots\ldots\ldots\ldots\ldots\ldots\ldots(12)$$

where SXES is saving on expenditure on exports,
SXUP is saving per unit of product, and
(YON - YNL) is reduction in crop yield per hectare.

6.10 A tax on nitrogen

So far, a limit on nitrogen use has been discussed on the basis of a restriction on the quantity used, as if a nitrogen quota could be readily implemented. An alternative approach is to increase the cost of nitrogen (CN) so that the economic optimum level (ON) is reduced to the specified nitrogen limit (NL).

The appropriate cost of nitrogen including tax (CNT) is derived from equation 9, by substituting NL for ON and CNT for CN, which gives:

$$CNT = V*(b - 2*c*NL) \quad \ldots\ldots\ldots\ldots\ldots\ldots\ldots\ldots\ldots\ldots\ldots\ldots (13)$$

where CNT is the cost of nitrogen (including tax) which would reduce the economic optimum level of nitrogen (ON) to the specified nitrogen limit (NL).

The tax, at (CNT - CN) per kg of nitrogen, would apply to nitrogen used (= NL) and would be a charge on the farmer, additional to the potential loss of income from having reduced nitrogen use. The nitrogen tax per hectare levied on the farmer, denoted by NTLF, is estimated as:

$$NTLF = NL*(CNT - CN) \quad \ldots\ldots\ldots\ldots\ldots\ldots\ldots\ldots\ldots\ldots\ldots\ldots (14)$$

where NTLF is the nitrogen tax levied on the farmer.

This tax would be payable by the farmer, and the proceeds to the exchequer would be available for distribution. The actual loss of income to farmers would depend on policies for distributing this revenue.

6.11 Focus on groundwater

The model outlined above focusses on nitrate concentration in the groundwater, as this is the main source of nitrate pollution for potable water. Surface water pollution would be a concern where these waters are used for drinking and where there is insufficient dilution with water of low nitrate concentration.

Chapter 7

EMPIRICAL ESTIMATES OF THE IMPACT OF RESTRICTING NITROGEN
ON CROP PRODUCTION AND NITRATE CONCENTRATION

7.1 Introduction

Estimates of the impact of a nitrogen restriction, as outlined in Chapter 6, can be made where the relationship between nitrogen use and both yield and nitrate concentration in groundwater are established. A major limitation is the lack of information on such relationships. This is understandable as awareness of nitrate pollution is relatively recent, the processes involved are relatively complex and the outcome is specific to individual sites. While acknowledging limitations of information, it is appropriate to analyse the limited data available to see what insights can be gained. Such analysis can point up the sensitivity of results to model parameters and may also give pointers regarding the appropriate forms of public intervention.

7.2 Analysis using selected data for Germany

The quadratic model, outlined in Chapter 6 above, is commonly found in the literature on crop response and nitrogen leaching. de Haen[1] used the quadratic model for data from West Germany. His barley yield (dt/ha) response model was: $Y = 23.1 + 0.4644*N - 0.001433*N^2$ ....(15)

---

[1] de Haen, H. (1982) "Economic Aspects of Policies to Control Nitrate Contamination Resulting from Agricultural Production", European Review of Agricultural Economics, Vol 9.

Nitrate leaching, as vertical nitrate flow, was estimated in a cross-section regression to be a quadratic function of nitrogen fertilizer, as follows:

$$NO3 = 23.59 + 0.3237*N + 0.002202*N^2 \quad \ldots\ldots\ldots\ldots\ldots\ldots\ldots (16)$$

This increasing quadratic function is consistent with the model outlined above (equation 5), which assumed that nitrate leaching is a linear function of NA, the amount by which fertilizer nitrogen exceeded that in the harvested crop. It is of interest to check how well the estimated quadratic function for leaching conforms to a linear relationship between NO3 and NA. Taking the nitrogen content of harvested barley as 2.0 kg per 100 kg of yield[1], gives nitrogen available for leaching (NA) as:

NA = N - 2*Y, where k = 2 in equation 4 above.

Values for NO3 and NA were estimated for N ranging from 0 to 320, by increments of 20, (i.e. 0, 20, 40, ... 300, 320). A linear regression relationship was estimated between the estimated values of NO3 and NA. The regression results were:

Standard Error of NO3 Estimate	6.876900
R Squared	0.996028
No. of Observations	17
Degrees of Freedom	15
Constant	85.22193
NA Coefficient	1.024143
Standard Error of NA Coefficient	0.016698

---

[1] Cooke, G.W. (1985) "The Present Use and Efficiency of Fertilisers and their Future Potential in Agricultural Production Systems", Environment and Chemicals in Agriculture, (Ed.) Winteringham, F.P.W., Elsevier Applied Science Publishers, London, pp 163-206, Tables 1,2.

The linear relationship fitted very well as evidenced by a high "R Squared". This linear relationship is shown in Figure 7.1.

FIGURE 7.1 Relationship between NO3 and NA.
(Based on estimated values)

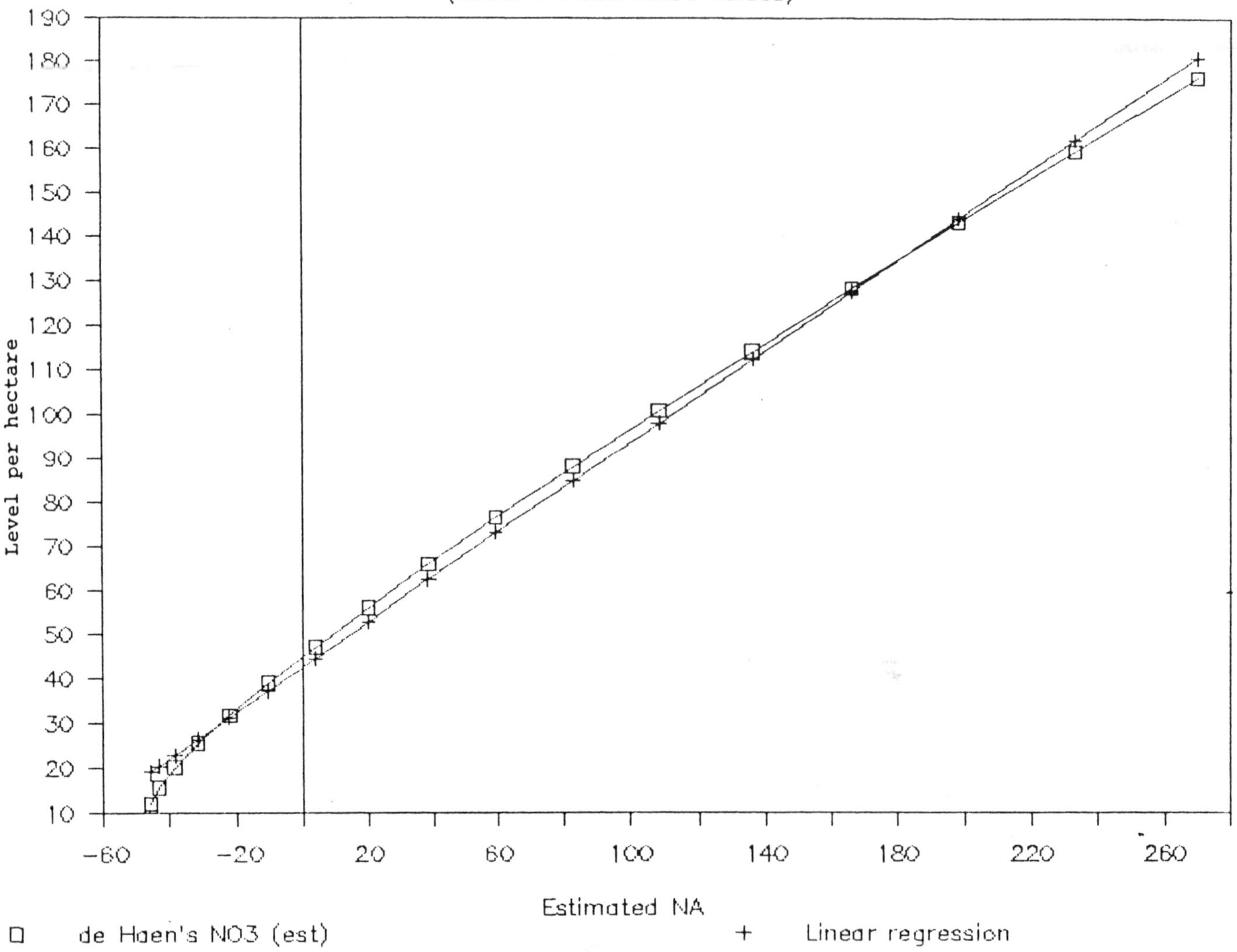

□ de Haen's NO3 (est)   + Linear regression

Also shown in Figure 7.1 are the corresponding estimates of nitrates in groundwater given by the quadratic function of nitrogen fertilizer. This indicates that a simple linear relationship may be appropriate, when nitrate leaching is related to fertilizer nitrogen in excess of that harvested in the crop. The parameters of such a linear relationship will vary with soil and climate. Soils which are rich in nitrogen and which facilitate mineralisation will have higher intercepts. Clay soils will have a less steep slope than sandy soils, as the latter facilitate drainage by their greater porosity. Climates with higher rainfall will have more dilute concentrations of nitrates in groundwater, which would be represented by a smaller intercept and a smaller slope for the linear relationship.

The estimated constant, (S = 85.2 for equation 5), is an estimate of what nitrate concentration would be if the land had been cultivated but no nitrogen applied and no crop grown. This indicates the importance of nitrogen released from the soil as a result of cultivation and mineralisation.

Even though the linear approximation was based on NO3 estimates given by equation 16, it gives different estimates of the coefficients for the quadratic relationship between NO3 and N. The linear approximation implies coefficients, as set out in equation 5, which are:

$$NO3 = 37.9065 + 0.0729*N + 0.00294*N^2 \quad \ldots\ldots\ldots\ldots\ldots\ldots\ldots\ldots(17).$$

These coefficients differ from those in equation 16 and hence equation 17 will give a somewhat different N value for any specified NO3. For example, 100 mg NO_3/litre derives from 126.3 kg/ha using equation 16 and 133.6 kg/ha using equation 17.

As outlined in the model, account has to be taken of the extent to which nitrates leached persist as nitrates in groundwater. de Haen suggests that persistence will be high "in the vicinity of pumping stations, at a high groundwater level and under good flowing conditions of the aquifer ...". He assumes an average persistence of 50%, which for the model outlined above gives:

$$NO3P = 0.5*NO3 \quad\quad\quad\quad\quad\quad\quad\quad\quad\quad\quad\quad (18)$$
where p = 0.5

Substituting de Haen's relationship for NO3, from equation 16, gives:

$$NO3P = 11.795 + 0.16185*N + 0.001101*N^2, \quad\quad\quad\quad (19)$$
where NO3P denotes mg NO_3 which persists per litre of groundwater.

Both quadratic functions — Y with a decreasing slope and NO3P with an increasing slope — are illustrated in Figure 7.2. The maximum barley yield of 60.73 dt/ha is associated with 162 kg N/ha and 66.8 mg NO_3/litre of groundwater.

In Figure 7.2 the nitrate concentration at zero nitrogen application is 11.8 mg per litre. Under arable cropping in sandy soils this intercept value is often much higher.

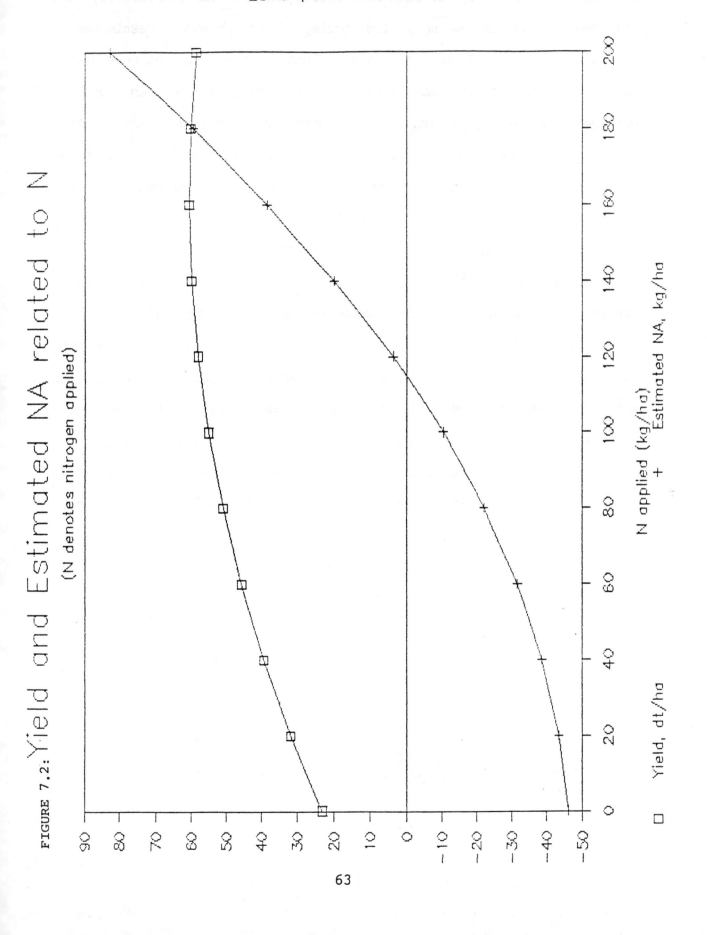

From data submitted by Pederson for Western Jutland (see Chapter 7.4), the intercept value is 30 mg/l for barley. Kolenbrander[1] estimates an intercept value of 40 mg/l for arable land if a 50 per cent persistence rate is assumed for nitrates leached. High nitrate leaching, at zero nitrogen application, arises from soil nitrogen which is mineralized. This mineralisation occurs mainly outside the period when nitrogen is being taken up by cereal crops. The nitrogen released is leached out in the autumn and winter, particularly when there is no crop cover, as with spring cereal crops. Then the mineralised nitrogen contributes very little to crop growth and would be represented in Figure 7.2 by an upward shift in the curve representing nitrate concentration. In Figure 7.2 the current maximum allowable concentration (MAC) of 50 mg per litre is reached when nitrogen is 127 kg per hectare, based on equation 19. If the intercept for nitrate concentration were 30, for example, 80 kg/ha of nitrogen would bring nitrate concentration up to the 50 mg/l limit. This highlights the importance of the intercept term and it is pertinent to note that it is relatively low in equation 19. The soil nitrogen level, which influences the intercept term, is largely determined by the history of land use.

7.2.1 Impact of nitrogen restriction on farmers' income

Assume that nitrogen use is restricted so as to attain a limit of 50 mg of nitrates per litre of groundwater. Setting NO3P in equation 19 equal to 50

---

[1] Commission of the European Communities (1978b) *The Spreading of Animal Excrement on Utilized Agricultural Areas of the Community - Summary and Conclusions*, Information on Agriculture, No. 51, Appendix I.

and solving for N gives the value of the nitrogen limit (NL), which is 126.8 kg N/ha. The potential loss in farm income (PLFI) from such a limit can be estimated for specified prices for barley and nitrogen.

The following average prices have been reported for Germany in 1986[1]:

barley price, V = 17.53 ECU/dt

nitrogen price, CN = 0.6899 ECU/kg

Where nitrogen is being considered as the only variable cost, these prices would indicate an optimum yield (YON) of 60.45 dt/ha, with nitrogen (ON) at 148.3 kg/ha. Limiting nitrogen to 126.8 kg/ha would reduce yield to 58.94 dt/ha, giving a potential loss (PLFI according to equation 10) of 11.67 ECU/ha.

If equation 17 had been used, rather than equation 16, the nitrogen limit would have been higher (NL = 133.6 kg/ha) and the potential loss in farm income lower (PLFI = 5.47 ECU/ha). This illustrates the sensitivity of the results to the curvature of the relationships in the region between the nitrogen limit and the optimum level of nitrogen. This curvature is determined by the coefficients of these relationships, (see equations 15 and 19).

Persistence rate for nitrates leached is another coefficient which influences the appropriate limit to nitrogen application rate. In the analysis above, half of the nitrates leached were expected to persist in the groundwater, (p = 0.5 in equation 18). If, for example, a 60 per cent

---

[1] Eurostat (1988) <u>Agriculture - Statistical Yearbook</u>.

persistence rate were appropriate it would lower the nitrogen limit necessary to ensure that nitrate concentration did not exceed 50 mg/l. The nitrogen limit would be reduced to 106 kg/ha rather than 127 kg/ha and the associated potential loss in farm income would be 43 ECU/ha as opposed to 12 ECU/ha. The impact on farm income is highly sensitive to the persistence rate, so it is important to establish an appropriate estimate of this parameter.

It has been pointed out above that the constant term (11.795) in equation 19 is quite low, so it is of interest to consider the impact of a higher value. If this constant were 30, the appropriate nitrogen limit would be 80 kg/ha rather than 127 kg/ha and the potential loss in farm income would be 117 ECU/ha rather than 12 ECU/ha.

The significance of reductions in income ranging from 12 to 117 ECU/ha depends on the levels of income from cereal production and the extent to which farms depend on cereals. Specialised "cereals type" farms in West Germany, according to 1985/86 FADN estimates, had 82 per cent of utilised agricultural area in cereals, with another 11 per cent in other field crops. The family farm income, before interest payments, was 170 ECU/ha on these farms. Ignoring rent paid on 46 per cent of the utilised area would bring this figure up to 240 ECU/ha, which is an estimate of income from cereal farming on owned land. Hence a reduction of 12 ECU/ha would be five per cent of income on owned land. If the constant term in equation 19 were 30, as it might be if soil nitrogen reserves were high, the reduction would be 117 ECU/ha or 49%. Where a restriction on nitrogen use reduced income from crop production, then rents would be likely to fall and this would offset at least part of the income decline. Hence the income decline per hectare on rented land would be less than that for owned land.

7.2.2 Possible savings at Community level

Limits on nitrogen use which reduce farm production would, in the case of surplus products, also reduce the expenditure involved in storing or exporting surpluses. Such savings for the Community could be set against the losses due to lower production. In 1986 export refunds for cereal exports averaged 6.22 ECU/dt exported. A reduction in yield of 1.51 dt/ha (Y_{ON} - Y_{NL} = 1.51, where NL is based on equation 16) would thus have been associated with a savings in export refunds of 9.41 ECU/ha. If these savings were given back to farmers it would reduce their potential loss of income to 2.25 ECU/ha. This figure, adjusted by the complete income multiplier (approximately = 2), would represent the loss to the EC as a whole, regardless of how savings on export refunds were distributed.

It was noted above that equation 17 would indicate a higher limit to nitrogen and hence a smaller reduction in yield. The corresponding saving in export refunds is (0.89*6.22 =) 5.54 ECU/ha, which is as large as the potential loss of farm income (in this case 5.47). This illustrates how less severe limits to nitrogen use will be offset to a greater extent by reduced EC expenditure on surpluses. This arises because the marginal return to nitrogen is lower at higher levels of use.

7.5 A tax on nitrogen

So far, a limit on nitrogen use has been discussed on the basis of a restriction on the quantity used. An alternative approach is to increase the cost of nitrogen (CN) so that the economic optimum rate of nitrogen application (ON) is reduced to the specified nitrogen limit (NL = 126.8 kg/ha). The appropriate cost of nitrogen including tax (CNT) is derived from equation 13. When this is evaluated, using equation 16 to estimate NL, the appropriate cost including tax (CNT) is 1.77 ECU/kg. This implies a tax rate of 157 per cent to achieve a reduction in nitrogen use of 14.5 per cent. The tax (at CNT - CN per kg of nitrogen) would apply to nitrogen used (= NL) and would be a charge on the farmer, additional to the potential loss of income from having reduced nitrogen use. This nitrogen tax amounted to 137.24 ECU/ha, which is very high relative to the potential loss of income from reduced nitrogen use, estimated at 9.39 ECU/ha. Since the rate of tax required to achieve the specified limit to nitrogen use is so high, some system of tax refund to producers would be necessary.

The revenue from the tax on nitrogen would be available to compensate producers, who could be given a refund of the tax for nitrogen used up to the specified limit (NL). Where limits on nitrate concentration are not surpassed, tax refunds could be given for all nitrogen used. While such a tax approach would work in principle it would involve massive administration in practice. It would require general monitoring of nitrogen use, even where its use did not give rise to environmental problems and would involve a massive task of administering tax collection and refunds. Inevitably the administration costs would mean less funds available for tax refunds. Furthermore it would be very difficult to enforce a two-tier price for nitrogen, as farmers could benefit by "illicit" trading between those entitled to a full tax refund and those who are not.

7.2.4 Sensitivity analysis

The analysis above is based on prices and export refunds in 1986. The sensitivity of the results to changes in these values was examined. In the period 1980 to 1986, the price of barley in West Germany was lowest in 1980 and highest in 1984. The effects of restricting nitrogen use to achieve 50 mg of nitrates per litre, using 1980, 1984 and 1986 prices and export refunds, are shown in Table 7.1.

Table 7.1: The effect of inter-year variation in prices and export refunds on the impact of restricting nitrogen use to attain a nitrate concentration of 50 mg per litre

Variables	1980	1984	1986
Cost of nitrogen in W. Germany (ECU/kg)	0.571	0.669	0.690
Barley price in W. Germany (ECU/dt)	16.460	20.040	17.530
Nitrogen/barley price ratio	0.035	0.033	0.039
Economic optimum for nitrogen (kg/ha)	149.93	150.39	148.31
Yield at optimum nitrogen (dt/ha)	60.52	60.53	60.46
Potential loss of farm income (ECU/ha)	12.67	16.04	11.67
Export refund rate (ECU/dt)	6.96	3.86	6.22
Cereal export refunds saved (ECU/ha)	10.95	6.14	9.42
Potential net loss of farm income (ECU/ha)	1.72	9.91	2.25
Cost of nitrogen including tax (ECU/kg)	1.66	2.03	1.77
Rate of tax on nitrogen (%)	191	203	156
Reduction from optimum nitrogen (%)*	15.5	15.7	14.5
Nitrogen tax levied (ECU/ha)	138.60	172.06	137.24

* The nitrogen restriction was set at 126.8 kg/ha to ensure that nitrate concentration did not exceed 50 mg/l, based on equations 15 and 19.

The high barley price in 1984 was 22 per cent above the low of 1980, while the ratio of nitrogen to barley price was highest in 1986, when it was 18 per cent above the low value of 1984. This 18 per cent change in price ratio leads only to a 1.4 per cent change in the optimum rate of nitrogen use and a 1.3 per cent change in optimum yield. Clearly the optimum nitrogen use is not very sensitive to price changes. This is why a massive tax on nitrogen would be required to induce a significant reduction in its use.

Attaining a standard of 50 mg of nitrates per litre of groundwater required nitrogen use to be below its economic optimum by between 14.5 per cent (1986) and 15.7 per cent (1984). The corresponding yields were in the range 2.5 to 2.6 per cent below the economic optimum. The associated potential loss in farm income ranged between 12 and 16 ECU/ha. Allowing for savings in export refunds on cereals, in the range 6 to 11 ECU/ha, the potential net loss was in the range 2 to 10 ECU/ha. If the reduction were to be induced by a tax on nitrogen, the tax rate would be in the range 157 to 203 per cent, with tax being levied at between 137 and 172 ECU/ha. The year 1984 was associated with the highest losses, as it had a highest product price and lowest rate of export refund.

While the MAC for nitrate concentration in potable water is 50 mg/l, the aim is to achieve levels of 25 mg/l or less. It is of interest therefore to examine the sensitivity of the impact on output and incomes to changes in the MAC. Table 7.2 shows the impact for limits of 50, 40 and 30 mg per litre, evaluated at 1986 prices and export refunds. As the limit is reduced, the impact of the reduction increases. Table 7.2 shows that changing the limit from 40 to 30 has a greater impact than changing from 50 to 40. The change from 40 to 30 requires a greater reduction in nitrogen use and the yield response to nitrogen is also greater at lower levels of use. The potential loss in farm income is quite sensitive to changes in the MAC, as the yield response to nitrogen increases as the rate of application is lowered. Changing the limit from 50 to 40 increased this potential loss by 41 ECU/ha, while a change from 40 to 30 increased it by 84 ECU/ha.

Table 7.2: The effect of changing the MAC evaluated at 1986 prices and export refunds*

Variables	Max. nitrate concentrations (mg NO_3/l)		
	50	40	30
Nitrogen limit (kg/ha)	126.76	102.6	74.6
Yield at nitrogen limit (dt/ha)	58.94	55.67	49.77
Potential loss of farm income (ECU/ha)	11.67	52.42	136.43
Cereal export refunds saved (ECU/ha)	9.42	29.78	66.45
Potential net loss of farm income (ECU/ha)	2.25	22.64	69.98
Cost of nitrogen including tax (ECU/kg)	1.77	2.99	4.39
Rate of tax on nitrogen (%)	157	333	537
Reduction from optimum nitrogen (%)	15	31	50
Nitrogen tax levied (ECU/ha)	137.24	235.53	276.27

* Based on equations 15 and 16.

If a tax on nitrogen were used to bring about the changes in optimum nitrogen use, the required increase in tax rate is greater at lower permitted nitrate levels. However, the tax levied per hectare increases less dramatically, as the higher tax rates apply to lower levels of nitrogen use. The analysis shows that potential loss in farm income is highly sensitive to the level of limit set for MAC.

Table 7.3 draws the above analysis together by showing the sensitivity to a number of technical parameters evaluated at 1986 prices. Using 1986 prices the optimum rate for nitrogen in Figure 7.3 was 148 kg/ha. The appropriate limit to nitrogen use is affected by the level of the MAC, nitrate concentration at zero nitrogen (i.e. the intercept for NO3P) and the persistence of leached nitrates. Estimates of potential income loss for different values of these parameters are shown in Table 7.3.

Table 7.3: Potential loss in farm income based on Figure 7.3 for a range of MACs, intercepts* and nitrate persistence rates.

@ 1986 Prices: optimum nitrogen = 148 kg/ha

Persistence	Intercept (mg NO_3/l)	MAC (mg NO_3/l)	N Limit (kg N/ha)	Potential loss in farm income	
				ecu/ha	%
60%	12	50	106	43	18
50%	12	50	127	12	5
		40	103	52	22
		30	75	136	57
	30	50	80	117	49

* Intercept for nitrate concentration indicates its level when nitrogen application is zero.

FIGURE 7.3: Barley yield & Nitrates in groundwater
Based on de Haen's yield response to N

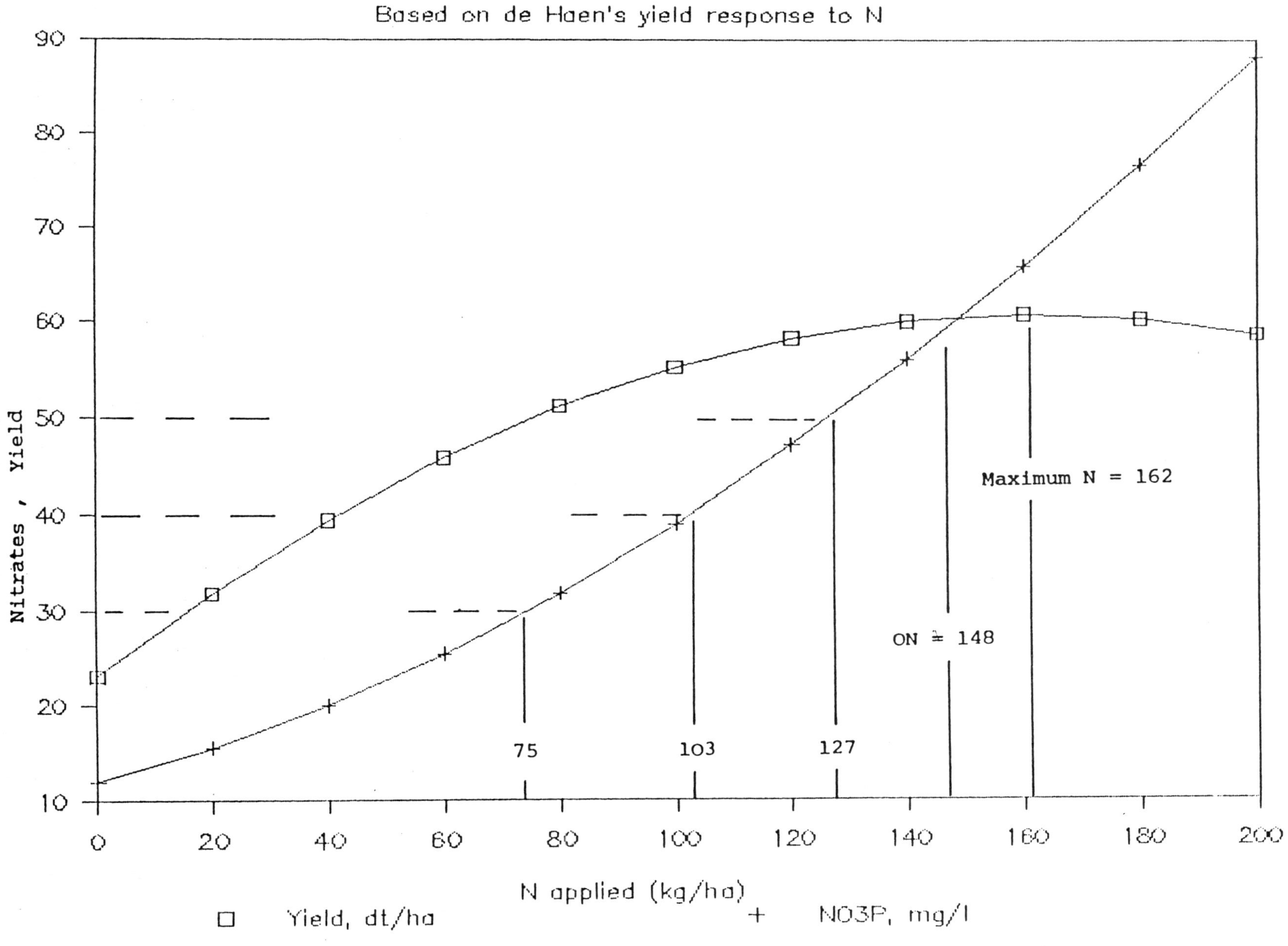

Table 7.3 shows that at 50 per cent persistence, a MAC of 50 mg NO_3/l would have required nitrogen to be reduced below its optimum level to 127 kg/ha, giving a potential income loss of 12 ECU/ha. The lower nitrogen limits to satisfy a MAC of 40 and 30 would lead to potential losses of 52 and 136 ECU/ha respectively. The potential income loss increases steeply as the MAC is lowered because the yield response to nitrogen is higher at lower rates of application. Hence, farmers have a vital interest in the level of MAC.

Soils with high nitrogen content, especially if cultivated, will have higher nitrate leaching and hence require even lower restrictions on nitrogen use. If, for example, the nitrate concentration at zero nitrogen in Figure 7.3 were 30 instead of 12 mg NO_3/l, the potential loss of income with MAC at 50 would be 117 ECU/ha rather than 12 ECU/ha.

A higher persistence of nitrates leached in the drainage water would also require lower limits on nitrogen use, to satisfy the MAC. If, for example, the persistence rate were 60 per cent as opposed to 50 per cent, this would have reduced the nitrogen limit from 127 to 106 kg N/ha and increased the potential income loss from 12 to 43 ECU/ha.

Clearly the required restriction on nitrogen use, and the associated income loss, are very sensitive to the parameters of the relationship between nitrate concentration and nitrogen application. These parameters vary between locations depending on soil/climatic conditions and the history of land use. The potential loss in farm income can be quite significant. The estimates above range from 12 up to 117 ECU/ha, which would be about 5% to 50% of the income on owned land.

7.3 Empirical results from Denmark

Dubgaard[1] analysed the effects on gross margin from crop production of a tax on nitrogen and a quota restriction on nitrogen use. These were based on a quadratic yield response to nitrogen. On heavy soils he estimated, using 1984 prices, that a 20 per cent tax on nitrogen would reduce its optimum use by about 5 per cent. A 100 per cent tax was associated with 16 to 26 per cent reductions in optimum nitrogen, depending on the crop rotation. The results indicate that the percentage reduction in gross margin is of a similar order of magnitude to the percentage reduction in optimum nitrogen use (see Table 7.4). Given the large impacts on gross margins, a tax refund (or a tax-free quota) for nitrogen use up to the specified limit seems appropriate. This result is consistent with the analysis based on de Haen's model outlined in Chapter 7.2.

---

[1] Dubgaard, A. (1987) "Reconciliation of Agricultural Policy and Environmental Interests in Denmark", Proceedings of the 11th Seminar of the European Association of Agricultural Economists, (Ed.) M. Merlo, G. Stellin, P. Harou and M. Whitby, Wissenschaftsverlag Vauk Kiel, pp 535-544.

Table 7.4: Impact of a nitrogen tax on optimum nitrogen rate and gross margin for crops, on heavy soils in Denmark

Cropping pattern	20% tax rate			100% tax rate		
	Reductions in			Reductions in		
	N rate (%)	Gross margin (%)	(ECU/ha)	N rate (%)	Gross margin (%)	(ECU/ha)
S. Barley only	6.1	10.0	17	26.1	30.9	74
W. Cereals/Rape[a]	4.9	5.4	24	20.2	24.5	108
W. Cereals/Peas[b]	4.4	3.2	17	15.9	14.9	76
Cereals/Sugar Beet[c]	6.0	3.3	20	18.8	15.0	90

[a] W. Barley, W. Oilseed Rape, W. Wheat, W. Wheat.
[b] W. Barley, Dried Peas, W. Wheat, W. Wheat.
[c] S. Barley, W. Wheat, W. Wheat, Sugar Beet.
Source: Dubgaard (1987) op. cit. Table 1.

Dubgaard also points out that a tax on nitrogen could have perverse effects, by increasing the relative profitability of nitrogen fixing crops, such as peas, beans and clover. More importantly he estimated that the time trend for nitrogen demand, exceeded the time trend for yields over the past thirty years. His demand model indicated that "...the very substantial fall in the relative price of nitrogen (relative to crop price) in the 1960s explains only about 10 per cent of the 70 per cent increase in the average nitrogen application rate in that period. Accordingly, structural and technological determinants seem to be the decisive factors behind the sustained growth in nitrogen application in Danish agriculture"[1]. If this pattern persists, then the reduction in nitrogen in response to a tax would be less than indicated in Table 7.4.

He examined the impact of restricting nitrogen use to 100 kg/ha, which

---

[1] Source: Dubgaard (1987) op. cit. p 537.

reduced nitrogen use to a similar extent as a 100 per cent tax. The loss to farmers' incomes, which varied by crop, was only 5 to 30 per cent of the loss under the taxation alternative. This also accords with the results in Table 7.1 above, where farm income loss under a restriction on nitrogen use was only eight per cent of that with a tax to achieve the same reduction on the optimum level of nitrogen. Dubgaard also notes that combining a 100 per cent tax on nitrogen with a tax-free quota of 100 kg/ha would allow farmers to exceed 100 kg/ha, where this was profitable. This could be achieved by taxing all nitrogen fertilizer and giving a refund for nitrogen used up to the limit of the farm quota.

The sensitivity to inter-year variation in prices was also examined by Dubgaard[1]. A decrease of 17 per cent in grain prices and 30 per cent in nitrogen price was forecast between 1985 and 1987. This was associated with about a five per cent increase in the optimum rate of nitrogen use. Combining the 1987 grain price with the 1985 fertilizer price would have reduced the optimum nitrogen rate by about five per cent. This illustrates again the relative insensitivity of the optimum rate of nitrogen application to prices.

Dubgaard also examined the impact of a reduction in product price on optimum nitrogen use. The ratio of nitrogen cost to product price determines the optimum nitrogen level. Hence a 20 per cent tax on nitrogen gives a reduction in optimum nitrogen similar to a 17.7 per cent decrease

---

[1] Dubgaard, A. (1986) Danish Agricultural Economy - autumn 1986, Institute of Agricultural Economics, Copenhagen, Table 16.

in product price. (Note that 100/120 = 83.3%). However, the impact of a product price decrease on farm income is much greater, as nitrogen costs were only a small fraction (under 10%) of the value of output. Reducing product prices does not seem an appropriate instrument for reducing nitrogen use, as its impact on farm income is very high relative to the impact on nitrogen use.

7.4 Pederson's data for Western Jutland region of Denmark[1]

Western Jutland (including the counties of Ribe and Ringkoebing) has sandy soil (>70% sand), an annual average of 755 mm rainfall and 415 mm of drainage water. Nitrate leaching is highly likely on such a soil, except where there is a permanent cover of a crop with a well developed root system. While the nitrate concentration in groundwater has not been measured, Pederson has provided "guesstimates" of the likely concentration in drainage water leached from the root zone, under different rates of nitrogen application. The extent to which this nitrate would persist, and eventually be found in the groundwater, is influenced by the soil. A general guideline for nitrate persistence is 25 per cent for heavy soils and 50 per cent for light soils. The soil in Western Jutland is very light so a 50 per cent nitrate persistence rate is the guideline for Pederson's data. Assuming a 50 per cent persistence rate Pederson's "guesstimates" of nitrate concentration in the root-zone can be converted to nitrate concentration in the groundwater.

---

[1] Pederson, Carl Age (1988) Private communication.

Permanent grassland with clover, which provides good crop cover and has a well developed root system, occupied nine per cent of the land. Table 7.5 gives the expected nitrate concentration in groundwater and the expected yields, for different rates of nitrogen application.

Table 7.5: Yield and nitrate concentration for different rates of nitrogen application to permanent grassland, in Western Jutland, Denmark

Grazing only:	Total N (kg/ha)[a]	150	75	0
	Yield (dt D.M./ha)	50	45	40
	Nitrates (mg/l)[b]	22	20	20
Grazing & harvesting:	Total N (kg/ha)[a]	200	100	0
	Yield (dt D.M./ha)	60	55	45
	Nitrates (mg/l)[b]	22	20	20
Zero grazing:	Total N (kg/ha)[a]	300	150	0
	Yield (dt D.M./ha)	70	60	50
	Nitrates (mg/l)[b]	22	20	20

a The highest rate is the recommended rate
b Assuming 50% of nitrates from the root zone persist.

The data indicate that permanent grassland ensures nitrate concentrations below the guideline of 25 mg per litre, which is half the MAC. These estimates point up an alternative approach to attaining adequate water quality. If the proportion of land devoted to such environmentally safe crops were high enough, then there would be adequate dilution of the high nitrate concentrations arising from other crops, when nitrogen is applied at the optimum rate. Restrictions might be placed on cropping mix so as to attain the desired water quality, assuming that nitrogen will be applied at the economically optimum rate. This is an alternative to controlling the use of nitrogen directly.

It should be noted that grassland could give rise to nitrate pollution of groundwater if total nitrogen application rates exceeded the recommended rates. This could arise from heavy application of animal manures, if grassland were being used as a disposal site for excess animal manures.

Ley grassland, (established less than 6 years), occupied 21 per cent of the land area of Western Jutland in 1986. Nitrate leaching is higher than for permanent grassland, which has a better developed root system and the recommended rate of nitrogen application is 50 kg/ha higher for ley grassland, (see Table 7.6). When nitrogen is applied at the recommended rate nitrate concentration is estimated at about 30 mg per litre. While this is above the 25 mg guideline, it is well below the MAC of 50 mg per litre.

Table 7.6: Yield and nitrate concentration for different rates of nitrogen application to ley grassland, in Western Jutland, Denmark

Grazing only:	Total N (kg/ha)[a]	200	100	0
	Yield (dt D.M./ha)	70	60	30
	Nitrates (mg/l)[b]	27	20	20
Grazing & harvesting:	Total N (kg/ha)[a]	250	125	0
	Yield (dt D.M./ha)	80	60	25
	Nitrates (mg/l)[b]	30	23	20
Zero grazing:	Total N (kg/ha)[a]	350	175	0
	Yield (dt D.M./ha)	90	60	20
	Nitrates (mg/l)[b]	33	25	20

a The highest rate is the recommended rate
b Assuming 50% of nitrates from the root zone persist.

With tillage crops the effect on nitrate leaching of reduced rates of nitrogen application depends on whether the reduction applies to both animal and chemical fertilizers or only to the latter. Reducing only chemical fertilizers gives less of a reduction in nitrate leaching. This is because "…. the fertilizer-caused nitrate leach-out is first and foremost due to the application of animal manure, especially when applied at an inopportune time". When nitrogen from all sources is reduced, Pederson's data indicate the typical quadratic-type relationship (of equation 5), with nitrate concentration being more responsive to a given reduction at higher rates of nitrogen application. When a reduction below the recommended rate is only in chemical fertilizer, and the animal manure application rate is maintained, then the reduction in nitrate leaching is lower. Table 7.7 gives the expected nitrate concentration when nitrogen reduction is applied to all nitrogen sources and not only to chemical sources.

Table 7.7: Expected nitrate concentration* for different rates of nitrogen application to tillage crops, under two systems of nitrogen reduction

Tillage crop		Slurry N	Total N relative to recommended rate		Zero
			Full	Half	
		(% of total)	(mg NO$_3$/l)		
Winter wheat:	A**	20	40.0	32.5	27.5
	B**	changing	40.0	37.5	27.5
Winter barley:	A	20	45.0	35.0	30.0
	B	changing	45.5	40.0	30.0
Spring barley:	A	20	45.0	35.0	30.0
	B	changing	45.0	40.0	30.0
Potatoes:	A	20	27.5	22.5	17.5
	B	changing	27.5	25.0	17.5
Fodder beet:	A	60	40.0	27.5	17.5
	B	changing	40.0	35.0	17.5
Spring rape:	A	30	50.0	35.0	27.5
	B	changing	50.0	45.0	27.5

* Assuming 50% persistence of nitrates found in the root zone.
** "A" denotes that nitrogen from all sources is being adjusted, so that the share from slurry is constant; "B" denotes that only chemical nitrogen is reduced in changing from full to half the recommended application rate.

Changing from full to half the recommended rate of nitrogen application gives a greater reduction in nitrate concentration than changing from half to zero. This is the usual pattern, (see Figure 7.2), where nitrate concentration changes more at higher rates of nitrogen application, for any given change in the rate of application. Where some of the nitrogen is from animal slurry and only chemical nitrogen is reduced, there is a smaller reduction in nitrate concentration.

The control of nitrate leaching is more problematic when animal manures are used as a source of nitrogen for tillage crops. A higher share of nitrogen from animal manures is associated with higher nitrate leaching on arable crops (equation 6). In vulnerable zones where animal manures are used on tillage crops, a limit in nitrogenous fertilizers would lead mainly to reductions in chemical fertilizers. In such circumstances the limit on total nitrogen use would need to be lower, or be supplemented by a code of good practice in relation to the use of animal manures, in order to attain a given rate of nitrate leaching. Hence the use of animal manures would complicate the implementation of controls on total nitrogen use.

At zero nitrogen the expected nitrate concentration from tillage crops (see Table 7.7) exceeds the 25 mg/l guideline, except for fodder beet. Tillage crops are a much greater threat to nitrate quality of water than grassland. It is to be expected that nitrate leaching from cereals, as well as from other arable crops, is greater than from grassland. Steenvoorden et al[1] found that ".... at the same fertilization level the nitrate concentration in the shallow groundwater is higher for arable land than for grassland. This is caused by mineralisation of the organic matter of the remainder of crop and roots on arable land and, moreover, by the absence in early spring of a growing crop and roots that can take up the mineralized nitrate."

Even at the recommended rate of nitrogen application, the expected nitrate concentration is in the range 40 to 50 mg per litre. Hence, it does not

---

[1] Steenvoorden, J.H.A.M. and Oosterom, H.P. (1979) Natural and Artificial Sources of Nitrogen and Phosphate Pollution of Waters in the Netherlands Surface, Technical Bulletin 114, Institute for Land and Water Management Research, Wageningen, The Netherlands, p 5.

exceed the current allowable limit of 50 mg per litre. These levels of nitrate concentration could be a problem if the MAC were reduced or if the nitrate concentrations increased over time. It should be noted that the data in Table 7.7 assumes a persistence rate of 50 per cent for nitrates, which is a general approximation. If higher persistence rates were appropriate for this sandy soil, then nitrate concentration would be expected to exceed the current MAC. It is of interest, therefore, to examine the implications of higher persistence rates and the sensitivity to changes in the persistence rates.

Cereal crops, which occupied 46 per cent of the land in 1986, are a potential source of nitrate pollution and spring barley accounted for 80 per cent of cereals. Assuming nitrate persistence rates of 60 per cent or more, nitrogen use would have to be restricted to below the recommended rate to keep nitrate concentration below 50 mg per litre. The impact of such a restriction was assessed, assuming that nitrogen from all sources (animal and chemical) was reduced. This assumption requires less restriction on nitrogen use than assuming that all the reduction is in chemical fertilizer.

A quadratic model for yield and for nitrate concentration was fitted through the three data points given in Table 7.8. The parameters for these quadratic equations are:

Yield (equation 1): $Y = 20 + 0.3846*N - 0.001183*N^2$

Nitrates (equation 5): $NO3 = 60 + 0.0769*N + 0.001183*N^2$

Table 7.8: Yield and nitrate concentration for different rates of nitrogen (20% from slurry) application to spring barley in Western Jutland, Denmark

Total nitrogen (kg/ha)*		130	65	0
Yield (dt/ha)		50	40	20
Nitrates: assuming 100% persistence (mg/l)		90	70	60

* The highest rate is the recommended rate

The yield equation implies a maximum of 51.26 dt/ha when nitrogen application is 163 kg/ha. The impact of restricting nitrogen is set out in Table 7.9, using 1986 prices in Denmark, when nitrogen cost 0.671 ECU/kg and barley was 17.59 ECU/dt. EC export refunds on cereals were 6.22 ECU/dt in that year. Under these prices the optimum nitrogen is 146 kg/ha and the optimum yield is 50.95 dt/ha, which is close to the yield of 50 dt/ha for the recommended rate of nitrogen use. Table 7.9 shows the effect of restricting nitrogen use to attain the limit of 50 mg per litre for nitrogen concentration in the groundwater.

Table 7.9: The impact of restricting nitrogen use* on spring barley, in the Western Jutland zone of Denmark, at two (60% and 65%) nitrate persistence rates

Variables	Nitrate persistence rate	
	60%	65%
Nitrogen limit (kg/ha)	112	91
Yield at nitrogen limit (dt/ha)	48.19	45.28
Potential loss of farm income (ECU/ha)	25	63
Cereal export refunds saved (ECU/ha)	17	35
Potential net loss of farm income (ECU/ha)	8	28
Cost of nitrogen including tax (ECU/kg)	2.12	2.96
Rate of tax on nitrogen (%)	216	341
Reduction from optimum nitrogen (%)	24	38
Nitrogen tax levied (ECU/ha)	162	209

* Nitrogen is restricted so that nitrate concentration does not exceed 50 mg/l and impact is evaluated at 1986 prices and export refunds.

Assuming 60 per cent nitrate persistence, nitrogen application would need to be reduced to 112 kg N/ha, which is 24 per cent below its optimum level of 146 N kg/ha. The potential loss in farm income would be 24 ECU/ha and the potential net loss, after allowing for export refunds saved, is estimated at eight ECU/ha. A fertilizer tax would have to be at 216 per cent to reduce the optimum nitrogen to 112 kg N/ha and tax levied would be 162 ECU/ha. To put these results in context it should be noted that FADN estimates for Denmark in 1985/86 give a family farm income, excluding rent and interest payments, of 190 ECU/ha on "cereals type" farms.

Table 7.9 also gives the estimated impact if nitrate persistence is assumed to be 65 per cent. This would require nitrogen use to be reduced to 91 kg/ha, which is 62 per cent of the optimum level. The associated potential loss in farm income is 63 ECU/ha and, after allowing for export refunds saved, the potential net loss is 28 ECU/ha. A change in the nitrate persistence rate from 60 to 65 per cent is associated with a 250 per cent increase in potential loss of farm income. The nitrate persistence rate, which is determined by the local soil/climate characteristics, has a big impact on the desired level of nitrogen use. This points up again the need to have nitrogen controls designed to suit local conditions, based on appropriate estimates of the technical relationships between nitrogen use and yield, nitrate leaching and nitrate persistence. "General guidelines" in relation to restrictions on nitrogen use would be inappropriate.

Offsetting nitrate pollution from tillage crops with ley or permanent grassland is possible if the grassland share of cropping is high enough. The appropriate cropping mix can be based on the assumption that nitrogen will be used at the rate recommended for the crop. The recommended rate is what is advised in advance of crop growth, while the economic optimum is not known until the crop is grown and weather and prices have been revealed. Recommendations take account of both price expectations and weather variability.

As an illustrative example, assume that nitrate persistence is 60 per cent and that permanent grassland is to be used to offset the spring barley crop, so as to bring the nitrate concentration in the groundwater down to the MAC 50 mg per litre. At recommended rates of application, spring barley would give a nitrate concentration of 54 mg NO_3/l (from Table 7.8) and permanent grassland 26 mg/l (Table 7.6 adjusted by 60/50 for 60% persistence rate). The area of permanent grassland (PGA) required to offset the spring barley area (SBA), to attain the limit of 50 mg NO_3/l is:

$$PGA = SBA*(54 - 50)/(50 - 26) = SBA*(1/6)$$

Based on the data provided, it is estimated that six hectares of spring barley could be offset by one hectare of permanent grassland. So, for example, a water catchment area which specialised in spring barley would have to have one in seven hectares under permanent grassland.

This approach would control nitrogen use indirectly by controlling cropping pattern. The relevant cropping pattern to control is that for a water catchment area. Hence it would be possible for excess nitrate leaching from some farms to be offset by low leaching from other farms in the same water catchment area. Restrictions on cropping pattern would apply only to farms contributing excess nitrates and only to the extent necessary to attain the desired water quality for the water catchment as a whole. Land cropping pattern would be much easier to monitor and control than nitrogen use. The technical information requirements would be less under this approach. Nitrate concentrations would need to be known only for nitrogen used at the recommended rates and observations under such rates would be more accessible, especially in vulnerable zones where nitrate concentrations were being monitored in any event.

It is pertinent to note that control of nitrate pollution is only one of the many environmental aspects that are of interest. These include other pollutants, (for example, phosphorous), and the preservation or creation of desirable ecological environments. The latter may be desirable for reasons varying from maintaining diversity of species to leisure and recreation uses of land. Control of cropping pattern could also be an instrument for achieving these wider environmental objectives and could form the basis of a more comprehensive approach to land-use management.

Increasing the share of crops which give lower nitrates in water would affect farm incomes. The overall loss in farm income could be estimated from the differences in income between crops. Where crop changes are being considered it would be necessary to take account of differences in overheads, as well as in gross margins. This is particularly so where cash crops are being replaced by grassland or other forage crops for animal feeding.

Non-cereal crops varied widely in their contribution to nitrate concentration. At the recommended rate of nitrogen application, the nitrate concentration (assuming 50% persistence) was only 28 mg/l for potatoes but was 40 and 50 mg/l respectively for sugar beet and spring sown rape. It is also interesting to note that the pea crop, which receives no nitrogen, is associated with a nitrate concentration of 37 mg/l, which is higher than for ley grassland receiving 250 kg of nitrogen per hectare.

7.5 Empirical results from the U.K.

England[1] estimated modified exponential production curves (in the form, $Y = a_1 - a_2(a_3^N) - a_4N$) and used them to assess the effect of a nitrogen tax on the optimum rate of nitrogen use and on farm gross margin. A 100 per cent tax on nitrogen reduced the optimum application rate for barley and wheat by nine to 13 per cent, depending on the crop and rotation. The reduction for linear programming model farms was similar (10% to 14%), except for a 24 per cent reduction where peas were substituted for rape as the cereal break crop. This shows how a tax on nitrogen could, in some regions, encourage expansion in nitrogen fixing crops, thus counteracting the loss in nitrogen intake from chemical fertilizers. These results confirm other findings that a tax on fertilizers is not effective in reducing nitrogen use relative to its impact on farm incomes.

---

[1] England, R. A. (1986) "Reducing the Nitrogen Input on Arable Farms", Journal of Agricultural Economics, Vol. 37, pp 13-24.

The 100 per cent rate of tax reduced the optimum yield by between one and three per cent, but the gross margin by between 159 and 246 ECU/ha. England also showed that a change in fertilizer price would have a much greater impact on nitrogen use than the reduction in grain prices which gives the same loss in gross margin.

7.6 Harvey's data for the Eastern Region of the UK[1]

Harvey provided data for the Eastern Region of the UK, which comprises Lincolnshire, Northamtonshire, Cambridgeshire, Norfolk, Suffolk, Bedfordshire, Essex, Hertfordshire and part of greater London. This region includes the majority of areas which are vulnerable to nitrate pollution in the UK, but data on specific vulnerable areas are not available. Data supplied for the region as a whole gave estimates of nitrogen lost (kg/ha) when nitrogen is applied at the recommended rate. There is a lack of data on both yield and leaching losses for nitrogen application rates below the recommended level. The lack of such data is a common and not unexpected problem. However it does indicate difficulties in drawing up guidelines for more restricted rates of nitrogen use.

All the nitrogen lost does not occur as nitrates. As a "rule of thumb" Bouwer[2] suggests that about half the nitrogen not taken up by the crop will

---

[1] Harvey, A. F. (1988), Environmental Protection Division, UK Ministry of Agriculture and Food, private communication.

[2] Bouwer, H. (1988) "Linkages with Groundwater", Nitrogen Management and Groundwater Protection, Elsevier, Amsterdam.

occur as nitrates in the drainage water, with the other half being denitrified. The average drainage water is given as 185 mm per annum, which is 1.85 million litres per hectare per annum. Nitrogen lost (kg/ha of N) is converted to nitrates in drainage water (mg/l) by multiplying by:

0.5 (to get share of nitrogen lost as nitrates),

1/1.85 (to allow for dilution), and

64/14 (to convert from nitrogen to nitrates).

This gives an overall multiplication factor of 1.23552. Allowance has also to be made for the persistence of nitrates in the drainage water. The region's soils, which are over limestone or chalk, are medium in the clay/sandy range (clay < 35%, sand < 70%). Using, as an approximation, a 25 per cent nitrate persistence for clay (>35% clay) soils and 50 per cent for sandy (>70% sand) soils, persistence rates in the range 30 to 40 per cent would seem appropriate for soils in the Eastern Region. The corresponding multiplication factors, to convert from nitrogen lost to nitrates in groundwater, would be 0.37 and 0.49. Using 0.4 and 0.5 as approximate conversion factors, Table 7.10 gives recommended nitrogen application rates for some crops in the region, along with nitrogen lost and expected nitrates in the groundwater.

Table 7.10: Nitrogen lost and expected nitrate concentration in groundwater when nitrogen is used at the recommended rates, for the Eastern Region of the UK

Crop	Nitrogen		Nitrate concentration	
	Recommended (kg/ha)	Lost (kg/ha)	NPR* = 30% (mg/l)	NPR = 40% (mg/l)
Grassland:				
grazing only	375	56	22	28
zero grazing	380	38	15	19
Winter wheat	225	50	20	25
Winter barley	160	24	10	12
Potatoes (maincrop)	220	110	44	55
Sugar beet	125	63	25	32
Winter rape	250	125	50	62

* NPR denotes nitrate persistence rate

The nitrogen lost varies from a low of 10 to 15 per cent for grassland and winter barley up to 50 per cent of recommended nitrogen rate for potatoes, sugar beet and winter rape. Table 7.9 shows a nitrate concentration of 50 mg per litre or less for all the crops, when 30 per cent nitrate persistence is used. Using a nitrate persistence of 40 per cent, only potatoes and sugar beet have nitrate concentrations exceeding 50 mg per litre. These two crops accounted for only 9.2 per cent of the land area in the region in 1986. These data would indicate that in general the Eastern Region does not have a nitrate pollution problem.

However, Murphy[1] asserts that "... in many arable areas in the Eastern Counties the nitrate level in water is now close to 100 mg per litre; twice the recommended level". (He estimates that reducing nitrogen use from its

---

[1] Murphy, M.C. (1987) The Value of Agricultural Land - Retrospect and Prospect, University of Cambridge.

current level of 185 kg/ha to 100 kg/ha would reduce yield by about 10 per cent and for a typical 160 hectare cereal farm this would reduce net farm income by 30 per cent, from stg£20,000 to £stg 14,000). Data for the region as a whole cannot provide appropriate guidelines for the more vulnerable zones within that region. This highlights the need to have guidelines, for controlling nitrate content of water, based on the local conditions affecting individual water catchments.

7.7 Limits to animal manures

In earlier sections of this chapter, we explored the limits to total nitrogen which would be necessary to ensure that nitrate concentrations in groundwater was under specified levels. As regards animal manures we have to consider not only the nitrogen, but also the phosphorous and potassium content, when setting limits to application rates.

Phosphorous is of particular concern as it can be a pollutant. In the Netherlands, control of animal manure applications is based on limiting the levels of phosphorous applied. Titulaer[1] (1988, p 361) sets out the Dutch government time-table for limits to applications of phosphate, which is currently set at 250 kg/ha of P_2O_5 for grassland but is to be progressively

---

[1] Titulaer, H.H.H. (1988) "Use of Organic Residues in Arable Farming", Agricultural Waste Management and Environmental Protection, Proceedings of the 4th International Symposium of CIEC, Vol. 2, (Ed.) Welte, E. and Szabolcs, I., Federal Agricultural Research Centre (FAL), Braunschweig-voelkenrode.

lowered to 125 kg/ha by the year 2000. The potassium based limit of 3.5 cow equivalents per hectare would give 140 kg P_2O_5/ha, so that the proposed phosphate limit of 125 kg/ha would imply a limit of 3.125 cow equivalents per hectare. On phosphate saturated soils, the Dutch regulations do not allow applications of phosphate to exceed that removed by the grass, which is about 110 kg/ha of P_2O_5. This is equivalent to cattle slurry from 2.75 cow equivalents per hectare. If cattle slurry is applied at higher rates it results in a build up of phosphates in the soil, phosphate leaching or both.

While a limit of 3.5 cow equivalents per hectare of grassland may be tolerable in the short run, it would have to be reduced to 2.75 in the longer run. On soils saturated with phosphate the limit would have to be 2.75 cow equivalents or less.

7.7.1 Limits to pig and poultry slurry on grassland

While "....with cattle slurry the main nutrient is often potassium, for pig or poultry-slurry it will usually be the phosphorous"[1]. The amount of pig and poultry slurry which would have 350 kg of K_2O would contain 410 and 729 kg of P_2O_5 respectively, as compared with 140 kg in the case of cattle

---

[1] Vetter, Heinz, Steffens and Gunter (1988) "Guidelines for an Economic Use of Slurry on Agricultural Land", Paper presented at the Joint FAO-subnetwork 4 and EC-cost and EC-COST Workshop, held at FAC Liebefeld, Bern, Switzerland, 18-22 June, p 6.

slurry[1]. Hence, phosphate is the nutrient which sets the limit to applications of pig and poultry manure. At the high current limit of 250 kg/ha of P_2O_5 set in the Netherlands, the nitrogen from pig and poultry slurry would be about 373 and 240 kg/ha respectively. The poultry slurry at this rate is not excessive. However, this rate of pig slurry gives too high a nitrogen rate, since animal manures usually account for only part of the total nitrogen application. A limit of 2.75 cow equivalents (245 kg) of nitrogen per hectare seems an appropriate limit, if the high standard (250 kg/ha P_2O_5) of phosphate is acceptable. However, the long term target in the Netherlands is half this rate. Crop uptake of phosphate (110 kg/ha P_2O_5) would justify a slightly lower rate.

7.7.2 Limits to animal manures on arable land

The current Dutch regulations set a limit of 125 kg/ha for P_2O_5 for all arable crops, except for green feed maize which will not have this limit until the year 2000. The corresponding level of nitrogen from slurry would be 278, 187 and 120 kg/ha for cattle, pig and poultry slurry respectively. Animal manures usually contribute only part of total nitrogen, as there is less flexibility regarding the timing of animal manure applications. The share of total nitrogen coming from animal manures tends to be low for cereals (25% or less) and higher for root crops, (up to 75%). Hence,

---

[1] Commission of the European Communities (1978a) The Spreading of Animal Excrement on Utilized Agricultural Areas of the Community, Volume I, Information on Agriculture, No. 47, Table 2.

regulation of phosphate applications will not necessarily translate into control of nitrogen applications. As a general guideline, a limit per hectare of about 1.2 cow equivalents (107 kg) of nitrogen for cereals and 1.5 (134 kg) for root crops seems appropriate.

Titulaer[1] gives an estimate of phosphate uptake by arable crops at 70 - 75 kg/ha of P_2O_5, which is two-thirds the rate for grassland. This is the appropriate limit for soils saturated with phosphate or if we wish to avoid a build up in the soil or phosphate leaching. The corresponding level of nitrogen from slurry would be 160, 108 and 70 kg/ha for cattle, pig and poultry slurry respectively. Except for cattle slurry these rates of nitrogen application are within the general guidelines for cereals given above. Since cattle are associated with green forage production, these strict phosphate limits would probably be adequate to prevent excessive nitrogen application from animal manures.

7.7.3 Summary

From the point of view of the control of pollution from animal manures, account must be taken of all farm animal sources as well as the constituent nutrients in manures. In relation to cattle slurry on grassland, the appropriate limit to avoid excess potash is 3.5 cow equivalents per hectare, but this would lead to a build up of phosphate and likely phosphate pollution. In order to avoid problems with phosphate, it is advisable to bring stocking density below 2.75 cow equivalents per hectare

---

[1] Titulaer (1988) op. cit. p 361.

of grassland. Where grassland is saturated with phosphate, stocking densities may need to be reduced below this level.

Pig and poultry slurry on grassland should be more restricted. An immediate limit of 2.75 cow equivalent of nitrogen (i.e. 245 kg N/ha) ought ultimately to be reduced to half this level to avoid a build up of phosphate in the soil.

Animal manures on arable crops need to be lower than on grassland, due to their lower uptake of phosphate and nitrogen. The share of total nitrogen coming from animal manures varies widely between crops, so that a lower limit is appropriate for cereals than for root crops. As a general guideline, a limit per hectare of about 1.2 cow equivalents of nitrogen for cereals and 1.5 for root crops seems appropriate. If manure applications were restricted to the limit of phosphate uptake by arable crops, this would probably be adequate to protect against nitrogen pollution from animal manures also.

It is noted that the Commission states in its "measures envisaged" that "In vulnerable zones, the rate of application of animal manures should be within the take up rate by crops"[1]. If this were applied to phosphates, it would probably avoid excessive nitrogen from animal manures.

---

[1] Commission of the European Communities (1988) Environment and Agriculture, Commission Communication, Com 338.

7.8 Implications of the empirical evidence

In principle, limits on nitrogen use could be set for individual crops so as to ensure that nitrate concentrations in groundwater were below prescribed limits and the trade-off between lower limits and greater potential income loss for farmers and for the EC can be estimated. The estimated impacts on production and income are quite insensitive to price changes. The percentage reduction in optimum nitrogen use induced by a tax is only one tenth of the tax rate, (see Tables 7.2 and 7.9). Hence, a tax on fertilizers is an ineffective instrument for reducing nitrogen use, while having a relatively large negative impact on farm income unless a system of tax refunds is also introduced.

The appropriate nitrogen restriction, along with its impacts, are quite sensitive to changes in the parameters of the underlying technical relationships, between nitrogen use and production, nitrate leaching and nitrate persistence. These relationships are site specific, so that separate estimates are required for each problem site, pending the development of general models of agricultural non-point source pollution[1]. The nitrate leaching relationship is complicated by nitrogen from animal manures, which are more problematic regarding the timing and control of application. The current lack of technical information is due in part to the fact that nitrate pollution is of relatively recent concern. This lack

---

[1] Giorgini, A. and Zingales, F. (Eds.) (1986) Agricultural Nonpoint Source Pollution: Model Selection and Application, Elsevier, Amsterdam.

of information and of established relationships points up the need for research in this area.

Better understanding of the technical relationships, and estimates of their coefficients, would also provide a basis for educating farmers on environmentally sensitive farming. Such education is necessary to create an awareness of nitrate pollution and to secure the adoption of environmentally friendly farming practices. Voluntary co-operation from farmers could make a contribution to reducing nitrate pollution, while compulsory quota-type restrictions on nitrogen use would be very difficult to enforce.

An alternative approach is to control nitrogen use indirectly by controlling the cropping pattern for a water catchment area. Excess nitrate leaching from some farms would be offset by low leaching from other farms in the same water catchment area. Restrictions on cropping pattern would apply to farms contributing excess nitrates and only to the extent necessary to attain the desired water quality for the water catchment as a whole. These farms could be assumed to apply nitrogen at the recommended rates, so the technical information requirements need only be established for these rates of nitrogen use. Observations on nitrate leaching under such rates would be more accessible, especially in vulnerable zones where nitrate concentrations would be monitored under current farming practice. Monitoring and controlling the cropping pattern of land-use would be easier than controlling the use of nitrogenous fertilizers. Controlling the pattern of land use would also be appropriate to achieving wider objectives for the rural landscape and environment and for land use policy in general. It would also be consistent with the promotion of "good farming practices", aimed at improving the utilisation of nitrogen by crops and reducing the amount of leaching.

Chapter 8

POLICIES AND MEASURES TO CONTROL NITROGEN POLLUTION IN MEMBER STATES

This section is based on submissions of EC Member States to the "European Study Days in Water Pollution and Intensive Farming", CEPFAR, Brussels, March 23-25, 1988. An extensive summary of each Member State's submission relating to this section is contained in APPENDIX 8 of the report. Opinions or comments contained in the appendix summaries reflect those of the Member State delegations concerned and do not necessarily represent the views of the study team.

Quite an array of policies and regulatory measures exist in individual Member States which aim to restrict pollution of water sources arising out of intensive agricultural practices. As yet there is no Community-wide legislation in this area. Policies adopted in Member States reflect the accepted seriousness of nitrate pollution and environmental protection needs in general. Consequently, policy initiatives are more advanced and restrictive on farmers where these considerations are high on the national political agenda.

Direct national legislative control of maximum chemical N fertilizer use by farmers does not exist apart from regulations in Belgium's Flemish region. On the other hand, regulatory requirements on farmers regarding animal manures are far more extensive throughout the Community. These include direct limits on farm animal populations per unit area and rules regarding the treatment, storage and spreading of animal manures. Restrictions particularly with regard to intensive indoor livestock enterprises exist in all Member States but to considerably varying degrees. These restrictions derive from legislation and other statutory orders for the protection of groundwater, general environmental concerns for rivers, urban dwelling areas, sensitive ecological areas etc. and farm building erections under planning laws.

Generally, restrictions relating to animal manures are imposed by local or regional authorities and hence diversity of measures can exist in individual countries which would reflect local or regional groundwater concerns, agricultural intensity, soil and climatic factors as well as politico-environmental considerations. Generally speaking, farmers are not exempted from regulations protecting rivers and lakes from organic pollution but enforcement in the agricultural sector has, in the past, tended to be weaker than for other sectors of the economy. Nevertheless, national governments are beginning to employ the "polluter pays principle" to farmers while at the same time the scope of environmental protection is being widened most notably to include groundwater sources of drinking water.

A number of general issues arise out of current policy measures related to animal manures, other farm practices and nitrogen pollution control. These include:

1. Cost of farmers' investments in enlarged manure storage capacity and the possibility of some farmers being forced out of business as a result.

2. Implications for the pattern of agriculture and farm incomes as a result of required farm practices e.g. autumn "green cover" in Denmark, maximum manure production per unit area in the Netherlands, animal stocking rate limits etc.
3. Manure depots and systems for transportation of manures from surplus to deficit areas.
4. Technical solutions to national manure surpluses are not imminent.
5. Member States prefer to use preventative and other voluntary measures to control nitrate contamination to agricultural activities. Major initiatives on appropriate research, farmer education and advice, and the adoption of good agricultural practice are advocated. While the voluntary approach is generally favoured, the introduction or extension of regulatory measures may follow if the voluntary approach is not successful.
6. Monitoring for compliance with regulations is acknowledged as a difficult administrative and fiscal problem. Technical support in terms of scientific standards and testing also need improvement.

On the basis of submissions to the aforementioned conference, the current position with regard to policy measures in the agricultural sector and the nitrates in drinking water at national level are summarised hereunder. Unless otherwise stated, direct national quantitative controls on N applications do not exist.

Belgium

(1) French speaking region:

> Good agricultural practice encouraged.

(2) Flemish region:

> 400 kg N/ha/per annum chemical fertilizers,
> Maximum stocking rate of 4 cow units/ha,
> Manure storage and spreading limitations,
> Limitation on size of intensive farms.

Denmark:

> Chemical N fertilizer use restricted in special cases involving nature reserves, sensitive groundwater etc.,
> General fertilization programme to be prepared by the farmer,
> Up to 65% of farmland to have vegetative cover until Oct. 20,
> Herd density restrictions,
> Manure storage and spreading limitations.

Federal Republic of Germany:

> Regional powers to restrict volume and timing of animal manure spreading,
> Regional restrictions on farm practices arising out of Federal laws for the protection of water sources and general environmental protection.

Greece:

> Rules regarding the installation of new farms,
> Treatment of animal wastes required by law,
> Rules on the location of farms and farm activities close to urban and other areas.

Spain:

> New legislation requires prior authorization for spreading animal manure and the payment by farmers of a manure levy,
> Some restriction on intensive dairy and pig farms and manures in the context of location, water protection and air pollution.

France:

> Chemical N can be restricted if required to protect water courses with provision for compensation to farmers in certain cases,
> Large intensive farm units must be authorised,
> Manure storage and spreading limitations.

Ireland:

>Planning permission for large farm buildings and associated facilities,
>Water pollution infringements punishable where recommended practices on manure storage and spreading are not followed.

Italy:

>No data were available to the study team regarding regulations relating to agricultural practices and nitrate abatement.

Luxembourg:

>Regulations are aimed at the protection of drinking water zones and in this context, some restrictions have been placed on local manure storage and spreading, particularly in the case of pig manure.

Netherlands:

>Regulatory control of manure practices are geared towards the protection of groundwater. National laws, based on phosphorus content in manures, are set for manure spreading and can be supplemented by local regulations. Farmers are obliged to keep a "fertilizer book" detailing the production and disposal own-produced animal manures.

United Kingdom:

>The voluntary code of good agricultural practice is relied upon as the principal means for ensuring environmentally safe practices relating to animal manures.

Chapter 9

PRINCIPLES AND POLICY OPTIONS

9.1 Introduction

The fundamental policy objective is to ensure adequately low levels of nitrates in drinking water. There are many approaches to meeting these objectives but they broadly fit into two principal strategic categories, namely policies to reduce the levels of nitrates getting into water supplies in the first place, and policies to reduce nitrates already in the water supply by treatment or other corrective methods. This study is concerned with the former approach.

The leaching of nitrates from farm land is a 'side-effect' of farming which, until relatively recently, was of little or no interest to either farmers or the general public. It was only when nitrate concentrations in potable water exceeded the guidelines set down for public health that leaching of nitrates into groundwater was perceived as a public nuisance. These groundwaters are part of the general environment which is a public good and when used as a source of drinking water give rise to public utility or well-being. Private production activities which damage this environmental utility imposes social costs (e.g. health risk, natural resource degradation) on others. Economic efficiency for society as a whole requires that production (and consumption) activities take account of the environmental costs which they impose. This is the basis of the "polluter pays principle" (PPP), which has been adopted by the EC Council[1].

9.2 Issues of political principle

Underlying the "Polluter pays principle" is the notion of a social contract between the citizen, in this case a farmer, and the wider society. Divergences of view regarding the appropriate basis for this social contract can range across a wide spectrum. At one end, is the view that land belongs to the landowner to do with as she/he sees fit, regardless of the effects on other members of society. At the other end of the spectrum, are people who would argue that farmers should be prohibited from land uses to which anybody else objects. A 'reasonable' social contract lies within these extremes, where the objections to land use are based on significant ill-effects on other members of society.

Social cooperation in restricting land use to keep nitrate concentration below the MAC, requires that farmers perceive the ill-effects of exceeding this limit to be significant. While the medical evidence in favour of any particular limit is beyond the scope of this study, it is important to recognise that its basis and acceptance are important to cooperation in enforcing limits.

Farmers' acceptance of restrictions, or of penalties for pollution, might be enhanced if they were also eligible for rewards for publicly desirable

---

[1] Commission of the European Communities (1975) *Official Journal*, No. L 194, July 1975, p 3.

'environmental products' of their farming. Thus it may be easier to promote good water quality as part of a wider programme of environmental enhancement, which might have possibilities of rewarding environmentally-friendly farming. Control of nitrate pollution from farming might be more feasible if it is seen as part of a more comprehensive environmental policy.

9.3 Practical problems with polluter pays principle

Nitrate pollution from farming is an effect which is external to the food production decisions of farmers. Hence farmers, have no reason to take such pollution into account when trying to optimise income earnings from farming. While such pollution remains external to them, farmers will tend to ignore the pollution consequences of their farming decisions. If farmers who caused nitrate pollution could be obliged to pay for the costs of water denitrification then these costs would be internalised and the polluter pays principle could be implemented. The polluter pays principle would require public authorities to impose a charge on polluting farmers, in accordance with their contribution to pollution, which would compensate society for the consequent resource degradation and costs of anti-pollution measures. The pollution charge would then become part of the farmers' production planning process. The principle is well established but its application can pose difficulties[1]. The application of this principle is particularly difficult with non-point source pollution, as in the case of nitrate pollution from farming.

---

[1] Pearce, D.W. (1976) Environmental Economics, Longman, London.

It is difficult to make polluting farmers pay for the pollution which they cause because responsibility for nitrate pollution of water cannot be clearly attributed to individual farmers. Nitrate pollution of groundwater was an unintended and, until recently, an unknown consequence of increased agricultural intensity. It is questionable whether farmers, who unknowingly contributed to nitrate pollution, can be held fully responsible for such pollution. Another complicating factor is the time it takes for nitrates leached from the soil to reach the groundwater. Much of the nitrates now in groundwater is a consequence of farming in an earlier period. This raises a problem as to how responsibility can be allocated between current and past "polluters". High nitrate concentration, or even a high reserve of soil nitrogen, caused by past farming practices may require more severe restrictions on current land use, which highlights the inter-temporal and dynamic aspects of environmental pollution. Even ignoring these aspects, it is difficult to attribute responsibility for current leaching of nitrates because nitrate pollution comes from diffuse, as distinct from point, sources.

Denitrification of groundwater water for drinking purposes is an expensive process[1]. Accordingly, attention is focussed on prevention of nitrate pollution arising from farming[2]. Hence, there is interest in getting farmers to farm their land in ways that will not lead to excessive nitrate

---

[1] HMSO (1986) <u>Nitrate in water</u>, Pollution Paper No. 26, HMSO London,

[2] Treatment (e.g. use of ion-exchangers) would require denitrification of only that portion of water used for drinking by people susceptible to methaemoglobinemia. However, the feasibility and costings of this approach need to be fully determined. Alternatively, clean drinking water could be supplied in bottles.

leaching. This means devising measures which will restrict farming practices, in particular nitrogen use, so as to avoid excessive nitrate leaching. These measures would require farmers, in areas where there is a risk of nitrate pollution, to conform to certain desirable or "good" farming practices, which would be tailored to local soil/climatic conditions. Farming practices may be controlled by regulating land use pattern, specific cultivation practices or fertilizer/manure applications to crops.

9.4 Voluntary restrictions on farming practices

Much of the nitrate pollution from farming is unintended and is often an unexpected or even unknown outcome. Education and persuasion have a role to play in (a) making producers aware of the environmental damage arising from certain farming practices and (b) encouraging the use of farming practices which are friendly toward the environment. If farmers were more aware of nitrate pollution risks, it would focus their attention on possible changes in their farming practices which would reduce the risks of such pollution. Environmentally careless farming may often reflect a lack of awareness and knowledge, rather than a lack of interest in the impact of farming on the environment. This is the case particularly in relation to animal manures. According to an OECD report[1], "..... animal fertilizers have tended to be considered as useless waste products and misused accordingly....adequate information, motivation and training of farmers

---

[1] OECD (1986) Water Pollution by Fertilizers and Pesticides, Paris, pp 15-16.

should be enhanced in member countries through 'Codes of Good Practices' for crop and animal production. Such Codes can be regarded as a key framework for both agricultural development and pollution prevention. They should take into account not only the optimum yield for the farmer, but also the essential soil, water and health protection criteria......Educational and advisory programmes based on 'Codes of Practice' are a convenient framework for pollution control... ".

Educating farmers about farming practices which are environmentally desirable is an essential element in controlling nitrate pollution, regardless of what other measures are taken. It is not surprising then that such an education programme is seen as a top priority and has found unanimous support from Member State delegations at the "European Study Days in Water Pollution and Intensive Farming", (CEPFAR, Brussels, March 23-25, 1988)[1]. The education of farmers in particular, and the public in general, can facilitate the development of understanding, cooperation and consensus among the different interest groups. While there is agreement on promoting "codes of good practice", the implementation of such codes requires (a) an acceptable basis for justifying the recommended practices and (b) measures to encourage their adoption. If "Codes of Practice" for farming are "...to prove fully effective, they are likely to need support from the appropriate regulatory and economic instruments, to provide adequate credibility and permanent incentive"[2]. In other words, such codes must be ultimately enforced lest they remain on the shelf as desired

---

[1] See APPENDIX 8.

[2] OECD (1986) op. cit. p 16.

objectives or mere statements of intent.

9.5 The need for research

Codes of practice must, in the first instance, be based on currently available information. There is a good level of general information on the process of nitrate leaching from farm land, which can guide the development of a code of good farming practice. These relate to the effect of factors such as: the timing of fertilizer applications (for both chemical and animal manures); the importance of growing crops in the autumn/winter period and the contribution of soil organic matter to nitrate pollution under arable cropping. However, rates of nitrate leaching, and the associated nitrate concentrations in groundwater, are strongly influenced by local soil and climatic conditions. Hence, guidelines for good farming practices need to be tailored to local conditions. Thus, the size and extent of the local area, in so far as nitrate control is concerned, is determined by external factors and this may lead to administrative difficulties if locally specific controls are to be implemented.

There is a lack of information on the relationship between nitrogen use and nitrate pollution of groundwater, as pointed out in chapter 7. Information on this relationship and the factors which influence it are not readily available, as we found in the course of this study, even for areas where there is a risk that nitrates will exceed the MAC (50 mg/l). This indicates a clear need for research to improve understanding of the nitrate leaching process and quantification of the relationship between nitrogen use on farms and both crop yield and nitrate pollution of groundwater, particularly in vulnerable areas. The development and parameterisation of quantitative models (such as the GLEAMS[1] model currently being developed[2] in the USA) seem an appropriate framework for this research[3]. Simulation models are needed for testing alternative farming systems and the technique of multiple goal programming is appropriate for exploring trade-offs between farming and environmental objectives[4]. In relation to monitoring at farm level, analytical methods for estimating the nitrogen content of soil and crops grown also seem relevant[5].

---

[1] Leonard, R.A., Knisel, W.G. and Still, D.A. (1987) "GLEAMS: Groundwater Loading Effects of Agricultural Management Systems", Transactions of the ASEA, Vol. 30, No 5, pp 1403-1418.

[2] GLEAMS is a model for vertical flux and the nutrient component of the model will be finished by January 1989, according to a private communication from W.G. Knisel, USDA-ERS.

[3] Giorgini, A. and Zingales, F. (Eds.) (1986) Agricultural Non-point Source Pollution: Model Selection and Application, Elsevier, Amsterdam.

[4] de Wit, C.T., van Keulen, H., Seligman, N.G. and Spharim, I. (1988) "Application of Interactive Multiple Goal Programming Techniques for Analysis and Planning of Regional Agricultural Development", Agricultural Systems, Vol ??

[5] Cooke, G.W. (1985) "The Present Use and Efficiency of Fertilisers and their Future Potential in Agricultural Production Systems", Environment and Chemicals in Agriculture, (Ed.) Winteringham, F.P.W., Elsevier Applied Science Publishers, London, pp 163-206

9.6 Integrating agricultural and environmental policy

While there seems to be general agreement on adopting a "code of good practice", it may not be adequate to rely on voluntary implementation of such practices, particularly where conforming to the code would adversely affect income from farming. The "code of good practice" ought to be supported by regulatory measures and economic incentives, to encourage its adoption as farming practice. The purpose of regulations embodying sanctions for failure to conform, or economic incentives, is to get farmers to incorporate into their decision-making processes the pollution consequences of certain farm practices. This indicates the need to integrate agricultural and environmental policy, so as to have a coherent strategy on land use.

In this broader context, policy measures ought to be assessed in relation to a number of objectives. These include objectives arising from the CAP, such as the protection of farmers' standard of living and technical and economic efficiency in food production, as well as the promotion of good farming practices to ensure that nitrates in water are at least within the MAC limit. The goal, in terms of policy measures, is a politically agreed land use policy which takes account of the diversity of land use and natural resource objectives.

In trying to arrive at appropriate constraints on or incentives for land use in farming, particularly in relation to nitrogen use, a number of issues have to be addressed. It has already been pointed out that the polluter pays principle is not directly operable in relation to nitrate pollution of groundwater. Where nitrogen use has to be restricted to below the economic optimum level to ensure acceptable water quality, the gain in water quality will lead to a loss in income from farming (see Chapter 7). Who should bear the losses arising from newly imposed restrictions?

It is pertinent to recall that current high concentrations of nitrates in water are due to past farming practices. Who benefitted from the past freedom to use land in a way that, often unwittingly, polluted groundwater with nitrates? Suppliers of nitrogen would have benefitted from profits arising from higher sales of nitrogenous fertilizers. Farmers would have benefitted from higher yields and incomes arising from the use of these fertilizers. The consequent increased supply of food might have been expected to benefit consumers through lower prices, but price support policies gave rise to increased costs associated with storage and export of surpluses. The main beneficiaries of the past 'freedom to pollute with nitrates' were farmers and suppliers of nitrogenous fertilizers. In so far as farmers were unaware that they were contributing to pollution, then the pollution might be more appropriately viewed as unintentional or accidental.

However, the situation regarding responsibility changes when people are aware that their farming practices do cause a level of nitrate pollution that is unacceptable to society as a whole. In particular, if restrictions are set for farming practice so as to avoid pollution, then it is clear what farmers can do without contributing to pollution. Farmers might reasonably be expected to operate within constraints relating to nitrogen and land use, which are legitimately introduced to meet health standards. While accepting that farmers should conform to approved farming practices, should they bear the full costs of any associated loss of income or be fully compensated? To indefinitely compensate farmers for farming so as not to pollute water supplies, would be to imply that they had a right to continue polluting but were yielding up that right in exchange for compensation. However it should be borne in mind that the introduction of constraints on farming practices would, by altering the profitability of farming, reduce the return on investments made before the environmental constraints were anticipated. It seems reasonable that farmers should be assisted for a time to adjust to the 'change in the rules' governing farming.

Investment in facilities to store and handle animal manures, so as to facilitate environmentally desirable manuring practices, should be eligible for assistance under prevailing schemes to aid farm development. Other environmentally desirable changes in farming systems, which require investment (for example changing from arable cash crops to permanent pasture) might likewise qualify for farm development aid. Additional compensation for income loss due to environmental constraints could be justified over a transitional period. In some situations income loss in food production might be offset to some extent by adapting the farming system to provide environmental goods or services, (such as game habitats; ecologically or aesthetically desirable landscapes). The possibilities for distribution of savings on the disposal of surplus commodities can alleviate further the income losses sustained, without necessarily increasing total public spending.

9.7 Regulatory measures and economic incentives.

It has been argued above that there is an apparent consensus for farmer education and the development of codes of good farming practice designed to avoid nitrate pollution from farming. These codes would be designed to improve the efficiency of nitrogen (both chemical and manure) uptake by crops. In particular they would indicate appropriate cultivation techniques and cropping rotations to reduce leaching of mineralised soil nitrogen. Codes of good practice are likely to need support from appropriate regulatory and/or economic instruments to provide adequate incentive for their effective implementation in farming practice. Farming practices may be controlled by regulating land use pattern or the rates of fertilizer/manure applications to crops. Since nitrate leaching is strongly influenced by local soil and climatic conditions, regulations would need to be tailored to local conditions and restrictions on nitrogen use would be required only in areas where there is a risk of nitrate pollution.

Regulating the amount of nitrogen applied to crops is one way of controlling the amount of nitrate leaching. In order to be fully effective it would be necessary to control the application of nitrogen from both chemical fertilizers and animal manures. The control of nitrate leaching is more problematic when animal manures are used on tillage crops, as this gives higher nitrate leaching. In vulnerable zones where animal manures are used on tillage crops, a limit in nitrogenous fertilizers would lead mainly to reductions in chemical fertilizers. The chemical nitrogen use permitted would have to take account of animal fertilizers applied, so that production, exchange and use of animal manures would also need to be monitored. An example of this type of control is the Danish requirement since 1987 that all farmers must establish fertilization programmes for their land[1].

The severity of nitrogen restriction needed to achieve satisfactory water quality will vary between locations. Arable farming in areas with high reserves of soil nitrogen would face particularly severe restrictions, as nitrate pollution problems tend to occur in such areas. In spring barley production, for example, it was shown in Chapter 7 that an appropriate restriction on nitrogen use could lead to a decline in income per hectare of 50 percent or even higher. The economic gain from ignoring such a nitrogen restriction could be a strong inducement to do so, even if farmers were at the same time being compensated for the expected loss in income arising from a nitrogen restriction. Hence some measures would be needed to ensure compliance with any nitrogen restriction which might be imposed.

One possibility is a farm quota restriction on chemical nitrogen, which would take account of animal manure use on the farm. However, as nitrogen restrictions are only relevant in some areas, transfers of quota-exempt purchases in other areas to farmers with quota restrictions could be profitable and result in the quotas being ineffective. Monitoring and

---

[1] See APPENDIX 8.

controlling such nitrogen transfers does not seem feasible, so a quota restriction seems inoperable.

At the scientific level, soil analysis for mineral nitrogen using the "N-min" method is widely used to estimate the available nitrogen in the soil in spring which can contribute to crop growth. Recent research in Germany[1] indicates that the "N-min" method can be used to determine the nitrate content in the soil in autumn at the start of the main leaching season. The cost-effectiveness of combining these and similar soil analysis techniques with programmes of financial compensation for lost income due to lower yields are worthy of investigation.

Plant analysis for nitrogen is a further possibility for monitoring and controlling, indirectly through crop payment penalties, the level of nitrogen applied to crops. While this approach may have potential for crops such as sugar beet and malting barley, difficulties would clearly arise in the case of crops where high protein content is required.

Another possibility is to reduce the optimum rate of nitrogen application by imposing a tax on chemical nitrogen. The rate of tax would have to be very high (see Chapter 7), about ten times the rate of reduction that it would induce in nitrogen use (e.g. 100% tax to induce a 10% reduction in crop output). Furthermore the appropriate reduction in nitrogen use, and

---

[1] Wiehmann, J. and Scharpf, H.C. (in press) "Reduction of Nitrate Leaching in a Vegetable Farm", Proceedings of EC Seminar Management Systems to Reduce Impact of Nitrates, September 24-25, 1988.

its associated tax rate, would differ between locations. Because the appropriate rates of tax would be so high they would have a very severe impact on farm incomes if applied to all chemical nitrogen used. (The change in product price to induce the same change in nitrogen use would have an even greater impact on farm incomes). Hence some system of 'nitrogen tax refunds' would be needed, which would present a large administrative task, including the monitoring of nitrogen use even in areas where nitrate pollution is not a problem. If the refund were closely related to nitrogen use it would defeat the purpose of the tax. If farmers were allocated tax-free quotas up to the appropriate nitrogen limit for the farm, it would solve the problem of income reduction but illicit trading of nitrogen quotas would be difficult to prohibit. It is concluded that a tax on chemical nitrogen fertilizers is not an appropriate mechanism for controlling the rate of nitrogen use.

A tax on nitrogen to finance the monitoring of nitrate concentrations in water and research into control of nitrate pollution may be justified. Such a tax may be justified on the grounds that use of potentially polluting nitrogenous fertilizers requires study of nitrate leaching and monitoring of the nitrate concentrations in water. This is a general requirement applicable to all areas, so a general tax on nitrogen would seem appropriate. It is pertinent to note here that suppliers and users of nitrogen have benefitted from freedom to use nitrogen in the past at rates which have already given rise to pollution.

Since nitrogen use is not amenable to direct control, an alternative approach is to regulate nitrogen use indirectly through the cropping pattern and livestock intensity on the farms. This approach would be complementary to the implementation of codes of good farming practice, which is the first element in a strategy to control nitrate pollution from farming. The codes of good practice, as outlined above, would be directed toward improving the efficiency of uptake of nitrogen by plants and reducing the nitrates in the soil at times of excess precipitation. At farm level this would involve aiming to have crop cover in autumn and winter, as well as adequate handling facilities for animal manures, to ensure their timely application at levels that do not pollute. This approach already raises the issues of appropriate crop rotations and livestock intensities.

The cropping pattern can control nitrate pollution as highly polluting crops such as spring cereals can be counterbalanced by a crop such as permanent grassland. If the proportion of land devoted to environmentally safe crops were high enough, then there would be adequate dilution of the high nitrate concentrations arising from other crops, even when nitrogen is applied at the optimum rate.

Control of total nitrogen use also requires limits to be set to livestock intensity, including livestock whose feed comes from outside the farm. The appropriate limit to livestock intensity would be that compatible with the cropping mix. The control of nitrate leaching is more problematic when animal manures are used, as they are associated with higher nitrate leaching, especially when applied to tillage crops. On farms with livestock, reductions in nitrogen use would probably be concentrated in the chemical nitrogen component. In such circumstances the limit on total nitrogen use would need to be lower, to allow for the increased share of nitrogen coming from animal manures. Farming according to codes of good practice is particularly important in relation to the use of animal manures, which are recognised as a major source of nitrate pollution.

Restrictions on cropping mix could be set so as to attain the desired water quality, assuming that nitrogen will be applied at the recommended rates. The recommended rate is what is advised in advance of crop growth and takes account of both price expectations and weather variability. Restrictions on land use and livestock intensity would have to be based on the technical relationships governing nitrogen and nitrate pollution of groundwater, which would take account of soil and climatic conditions. Nitrate concentrations need only be established for nitrogen used at the recommended rates and observations under such rates would be readily accessible, especially in vulnerable zones where nitrate concentrations were being monitored in any event. Land cropping pattern would be much easier to monitor and control than nitrogen use.

The relevant cropping pattern to control is that for a water catchment area. Excess nitrate leaching from some farms could be offset by low leaching from other farms in the same water catchment. Restrictions on cropping pattern need only affect farms contributing excess nitrates and only to the extent necessary to attain the desired water quality for the water catchment as a whole. Farms with low nitrate leaching would be free to increase nitrogen use and leaching, as long as they conformed to the restrictions set down for their location and water catchment.

It has been argued above that farmers might be compensated, over a transitional period, for income losses arising from adjustments to new environmental regulations. Whether conformity with these regulations gives rise to an income loss, as well as the level of this loss, would be judged by reference to the pattern of land use in a recent reference period. Compensation would apply only to those farmers whose land use pattern was adversely affected in relation to income generation. Those whose current farming conformed to the restrictions would not be eligible for compensation, while all farmers would be free to vary their land use pattern within the scope allowed by environmental restrictions. In some farming areas income losses from crop substitution might be greater than for reducing nitrogen use on the current crop mix. A reduction in nitrogen use could be accepted as an alternative to changing crop mix, where an acceptable procedure for monitoring nitrogen use is agreed with the authority implementing controls on nitrate pollution. Controlling the pattern of land use is consistent with the promotion of good farming practices and is also appropriate to achieving wider objectives for the rural landscape and environment and for land use policy in general.

It is pertinent to note that control of nitrate pollution is only one of the many environmental concerns. These include other pollutants, (for example phosphorous), and the preservation or creation of desirable ecological environments. The latter may be desirable for reasons varying from maintaining diversity of species to leisure and recreation uses of land. Control of cropping pattern could also be an instrument for achieving these wider environmental objectives and could form the basis of a more comprehensive approach to land-use management.

9.8 Implementation of controls on nitrate leaching from farming

In all EC Member States there are regulations governing the use of farmland. These usually take the form of prohibitions on certain practices, supported by legal sanctions. Some aspects of a code of good farming practice may already be governed by existing regulations. However, good practices in relation to avoidance of nitrate pollution, as well as other new and emerging concerns, are unlikely to be adequately dealt with by existing regulations. Where new regulations are being introduced or old ones changed, compensation for consequent losses may be justified. It has been argued, in relation to regulations on land use designed to avoid nitrate pollution, that compensation may be justified for a transitional period.

Restrictions on nitrogen use in farming have direct impacts on farm production and income, which are central concerns of the CAP. Reductions in production arising from lower use of nitrogen would reduce the costs of disposing of surplus farm production. It seems appropriate therefore that Community funds should make a contribution toward transitional compensation for loss of income arising from new restrictions and toward the costs of investment needed to implement the required farming practices. In relation to "limiting environmentally undesirable developments", EC Ministers for the Environment have concluded that "... the structural funds constitute a privileged instrument in achieving such aims, as well as for the integration of the environmental dimension in the agricultural sector"[1]. It also seems appropriate that regulations to control nitrate leaching from farming should be drawn up in consultation with the Commission if their implementation is to qualify for Community funding.

Controls on either nitrogen use or land use in farming need to be designed for the particular water catchments where nitrate pollution of groundwater is a problem or is likely to become a problem. This means that, while health standards are set at Community level, the design of regulations to achieve these standards will have to be established locally and implemented at farm level by a local authority. The local administrative arrangements for implementation are likely to vary, as they will have to be compatible with established institutional structures. The relationship between the implementing agency and farmers might be operated via "management agreements".

Management agreements are agreements to manage farms in a prescribed way so as to achieve some desired environmental objectives. The UK Countryside Commission gave the following definition in 1973:

> "A management agreement may be described as a formal written agreement between a public body and an owner of an interest in land (the term "owner" may here include lessees and occupiers) who thereby undertake

---

[1] Commission of the European Communities (1988) Environment and Agriculture, Commission Communication, COM 338, June 8.

to manage the land in a specific manner in order to satisfy a particular public need, usually in return for some form of consideration"[1].

In relation to control of nitrogen use, transitional compensation would be payable to those whose farm incomes were reduced by conforming to new nitrogen control regulations. Grant aid for capital investments needed to avoid nitrate pollution could also be incorporated in such agreements. When the terms of financial support are specified farmers might be invited to participate voluntarily in such management agreements, so as to encourage social cohesion in pursuit of environmental objectives. This would reduce, if not eliminate, the need for compulsory measures. In the Netherlands, the Provincial Committee for Land Management which consists of government officials and representatives of farmers' and nature conservation organizations, drafts a management plan for environmentally sensitive areas[2]. This type of structure seems appropriate to joint action to ensure water quality, among other environmental objectives.

The Commission, in its fourth programme of action on the environment stressed the need to integrate environmental and other policies, including the CAP and a recent EC workshop on environmental management in agriculture

---

[1] Commission of the European Communities (1987) Agriculture and the Environment: Management Agreements in four countries of the European Communities, Report for the Commission, p 13.

[2] de Boer, T.F. and Reyrink, L.A.F. (1988) "The Netherlands, II: Policy", Environmental Management in Agriculture - European Perspectives, (Ed.) Park, J.R., Belhaven Press, London.

concluded that "the watchword is integration"[1]. The management agreement seems a useful mechanism for integrating agricultural and environmental policy. Management agreements are being used to an increasing extent in some Member States and can be seen as a means of regulating land use planning so as achieve environmental objectives. These objectives can include diverse goals, ranging from resource (e.g.water) protection to the protection or creation of specific ecological systems. Farmers, by their food production systems, have a dominant influence on the rural environment. Public interest is shifting increasingly toward the environmental, including ecological, impact of food production.

Management agreements provide a mechanism whereby farmers could be paid for providing a publicly desirable rural environment. Farmers could not charge individual members of the public for the use of many environmental qualities and hence there is a role for agreements with public authorities. Where the public interest and farmers' private interests do not coincide, farmers can be offered incentives in exchange for a commitment to farm in a socially desirable way. Management agreements evolved initially as a means of protecting landscapes and wildlife habitats. These agreements are in accordance with EC Regulation 797/85 which, under Article 19, allows Member States "to take measures to introduce special national schemes in environmentally sensitive areas with the objective of maintaining farming practices which are compatible with the requirements of protecting the countryside and ensuring an adequate

---

[1] Park, J.R. (1988) <u>Environmental Management in Agriculture - European Perspectives</u>, Belhaven Press, London.

income for farmers". Now a wide variety of management agreements are found in some Member States[1]. While management agreements on controlling nitrogen use would have to be designed and monitored locally, they could be drawn up within a general framework agreed at Community level. This would accord with a recent Commission study on management agreements which suggests that "the best way forward might be the introduction of a flexible scheme negotiated at the national or regional level, but within a broad EC framework"[2].

While the establishment of overall guidelines concerning maximum total nitrogen use may be generally useful, they would have to be modified to local conditions such as soil nitrogen, animal manure production (see Chapter 7) and nitrate situation before such guidelines could become operational. Possible Community-wide guidelines in this regard might be 200 kg N/ha for cereals and 400 kg N/ha for grassland.

At the level of the nitrate vulnerable zone, more specific measures could be undertaken. These might include individual farm plans to ensure that the MAC is not exceeded. A possible method of policing such farm plans might entail the measurement of soil NO_3-N content in October in order to establish estimated plant uptake in the coming months and thereby the amount of nitrates available for leaching as referred to earlier. However, this approach may not be relevant to all crops and the establishment of revised product payment systems would require a major political initiative.

---

[1] Commission of the European Communities (1987) op. cit. pp 21-22.

[2] Ibid. p viii.

SUMMARY OF QUADRATIC MODEL OUTLINED IN CHAPTER 6

(Loose leaf for easy reference)

This summary gives the model equations and explains the notations used. Note that * where used in equations, **indicates multiplication**

$N = NE + SN$

where N is total nitrogen in kg/ha,
NE is the chemical nitrogen equivalent of animal manure, and
SN is chemical nitrogen in kg/ha.

$$Y = a + b*N - c*N^2 \quad \dots \dots \dots \dots \dots \dots \dots \dots \dots \dots \dots \dots \dots (1)$$

where Y is yield in dt/ha, and
a, b and c are technical coefficients.

$$NC = k*Y \quad \dots \dots \dots \dots \dots \dots \dots \dots \dots \dots \dots \dots \dots (2)$$

where NC is the nitrogen in the crop yield (kg per ha), and
k is the nitrogen content of the crop (kg N per 100 kg).

$$NB = \{(1 - k*b) + or - [(k*b - 1)^2 + 4*k^2*a*c]^{0.5}\}/2*k*c \quad \dots \dots \dots (3)$$

such that NB>0,
where NB is the rate of nitrogen application such that NC = N.

$$NA = - k*a - (k*b - 1)*N + k*c*N^2 \quad \dots \dots \dots \dots \dots \dots \dots \dots \dots (4)$$

Where NA is the nitrogen not taken up by the crop.

$NO3 = S + w*NA,$

where NO3 is nitrate leaching (mg/litre of drainage water),
S is NO_3 from sources other than nitrogen applied, and
w is a technical coefficient.

$$NO3 = d + e*N + f*N^2 \quad \dots \dots \dots \dots \dots \dots \dots \dots \dots \dots \dots \dots (5)$$

Where the technical coefficients are d = S - w*k*a,
e = w - w*k*b and
f = w*k*c.

$$NO3 = d + e*N + f*N^2 + g*(AN) \quad \dots \dots \dots \dots \dots \dots \dots \dots \dots \dots (6)$$

where AN is the total nitrogen per hectare from animal slurry, and
g is a technical coefficient appropriate to the animal slurry.

$$NO3P/p = d + e*N + f*N^2 \quad \text{...(7)}$$

where NO3P denotes nitrates that persist in the water, and
p is the portion of NO3 that persists.

$$NL = \{- e + or - [e^2 - 4*f*(d - DNO3)]^{0.5}\}/(2*f) \quad \text{....................(8)}$$

such that NL>0.
where NL is the upper limit to N which ensures that NO3P does
not exceed DNO3 -the desired upper limit for NO3P.

$$ON = (b - CN/V)/(2*c) \quad \text{...(9)}$$

where ON is the optimal level level of N,
V is the value per unit of the marginal Y produced, and
CN is the cost per unit of chemical fertilizer N.

$$PLFI = V*(YON - YNL) - CN*(ON - NL) \quad \text{...............................(10)}$$

where PLFI is potential loss of farm income per hectare of crop,
YON is yield with optimum nitrogen application (N = ON), and
YNL is yield with nitrogen at the limit rate (N = NL).

$$PLTI = M*(PLFI) \quad \text{...(11)}$$

where PLFI is the potential loss in farm income per hectare, and
M is the complete Moore-type income multiplier.

$$SXEX = SXUP*(YON - YNL) \quad \text{...(12)}$$

where SXES is saving on expenditure on exports,
SXUP is saving per unit of product, and
(YON - YNL) is reduction in crop yield per hectare.

$$CNT = V*(b - 2*c*NL) \quad \text{...(13)}$$

where CNT is the cost of nitrogen (including tax) which would reduce
the economic optimum level of nitrogen (ON) to the
specified nitrogen limit (NL).

$$NTLF = NL*(CNT - CN) \quad \text{...(14)}$$
where NTLF is the nitrogen tax levied on the farmer.

APPENDICES

(Appendix numbers relate to the relevent
chapters in the report)

APPENDIX to CHAPTER 1

Covering letter and question sheets circulated to delegates from Member States who attended the study days on "water protection and intensive farming", organised by CEPFAR and held in Brussels on March 23-25.

* * * * * * * *

To:

Re: DG VI Study P244 on:

"Intensive farming and the impact on the environment and rural economy of restrictions on the use of chemical and animal fertilizers"

ECN_QSH.APR   April 30, 1988

Dear

Enclosed find a copy of the three question sheets circulated to you during the study days (Brussels, March 23-25) on "water protection and intensive farming", organised by CEPFAR. The definition of "zone" has been changed from "zones where nitrates in water exceeds 40 mg/litre" to **"zones where nitrates in water are likely to be a problem"**.

It is hoped that relevant information may already exist for some zones (areas) which were perceived as being problematic. An objective of the study is to assess the impact of restricting N used in farming on (a) nitrates in drinking water and (b) on crop production and associated income from farming. Since the study is to be completed by next June, we have to rely on readily available data and hope you can assist in providing some data for vulnerable zones in your State.

Data requested on **question sheet 3** are **vital** for our study and can be **based on expert opinion**, regarding the **response to total N** (animal plus chemical) of both **yield and nitrate leaching**. Question **sheet 2** is the next most important, as it would give the **pattern of land-use and of recommended fertilizer use**, both animal and chemical. Question **sheet 1** seeks to establish some pertinent **soil and rainfall** data, **farm size structure and production systems**, as well as the **sources and farm-gate prices of chemical**

Nitrogenous fertilizers.

I appreciate that only some of the data requested may be available and only for some vulnerable zones in your State. In order to reveal the diversity of vulnerable zones in the EC it would assist if you could provide whatever data are available. Given the short duration of the study, it is important to let me have any available data as quickly as possible.

Please let me know immediately what data you can make available and relevant persons to contact.

Sincerely,

A G Conway
Study Director

c.c. A. Hardt, Secretary-General of CEPFAR
     Dr. A. Moreale, Head of Studies DG VI
     Dr. Nigel Robson, Head of Special Services DG VI

FARMING IN ZONES WHERE NITRATES IN DRINKING WATER IS LIKELY TO BE A PROBLEM - (SHEET 1)

ZONE:name:_____ EC Region_____ State:___ Nitrates(mg/l)____

SOIL No__ 1 >35% clay; 2 18-35% clay; 3 <18% clay & <70% sand; 4 >70% sand.

RAINFALL:(mm/an)_____ AVERAGE amount of DRAINAGE WATER:(mm/an)_____

	Total (1000)	Average no. by farm size (ha of UAA)					Period Housed mth/yr
Livestock in year: 19__		1-<5	5-<10	10-<20	20-<50	>=50	
Total cattle	____	____	____	____	____	____	____
of which: dairy cows	____	____	____	____	____	____	____
other cows	____	____	____	____	____	____	____
Total sheep	____	____	____	____	____	____	____
of which: breeding sheep	____	____	____	____	____	____	____
Total goats	____	____	____	____	____	____	____
of which: breeding goats	____	____	____	____	____	____	____
Total horses	____	____	____	____	____	____	____
Total pigs	____	____	____	____	____	____	____
of which: breeding pigs	____	____	____	____	____	____	____
Total poultry	____	____	____	____	____	____	____
of which: laying	____	____	____	____	____	____	____
Labour (AWU[1]) no.	____	____	____	____	____	____	
Land (UAA) ha	____	____	____	____	____	____	
Farms (total no.)	____	____	____	____	____	____	
Land use percentages:							
Grain cash crop %	____	____	____	____	____	____	
Root cash crop %	____	____	____	____	____	____	
Permanent crop %	____	____	____	____	____	____	
Green fodder %	____	____	____	____	____	____	
Root fodder %	____	____	____	____	____	____	
Grassland %	____	____	____	____	____	____	

Synthetic N fertiliser:

Sources	% from source	Farm-gate Price/kg N
1 Ammonium nitrate	_____	_____
2 Urea	_____	_____
3 Other (specify) _____	_____	_____

Main source code __ Code no. 1 Ammonium nitrate; 2 Urea; 3 other

---

[1] AWU denotes annual work units equivalent to full-time workers.

FARMING IN ZONES WHERE NITRATES IN DRINKING WATER IS LIKELY TO BE A PROBLEM - (SHEET 2)

ZONE:name:_____ EC Region_____ State:___

Land use in Year: 19__

	Area	Fertiliser for optimum yield			
		Cattle	Pig	Poultry	Synthetic N
	100 ha	t/ha	t/ha	t/ha	kg/ha
Total land area					
Woods and forests					
Rough grazing in use					
Good permanent pasture					
of which: grazing only					
grazing & harvesting					
zero grazing					
Total permanent crops					
of which: vineyards					
olives					
(specify) other major_____					
Ley grassland (under 6 years)					
of which: grazing only					
grazing & harvesting					
zero grazing					
Total green fodder					
of which: maize					
(specify) other major_____					
Total root fodder					
of which: beet					
(specify) other major_____					
Total cereals					
of which: durum wheat					
common winter wheat					
common spring wheat					
winter barley					
spring barley					
(specify) other major_____					
Grain maize					
Total root cash crops					
of which: potatoes					
sugar beet					
(specify) other major_____					
Total oilseed crops					
of which: sunflower					
rape					
(specify) other major_____					
Other cash crops					

FARMING IN ZONES WHERE NITRATES IN DRINKING WATER IS LIKELY TO BE A PROBLEM - (SHEET 3)

ZONE:name:_____ EC Region_____ State:___

RECOMMENDED TOTAL NITROGEN (including the synthetic fertiliser equivalent of animal manures) and the RESPONSE in terms of YIELD and NITRATE CONCENTRATION	Recommended total $N(=R)$ kg/ha	Expected yield (100kg/ha) for			Expected leaching (mg NO_3/l) for		
Land Use		N=R	N=R/2	N=0	N=R	N=R/2	N=0
Good permanent grassland							
of which: grazing only	___	___	___	___	___	___	___
grazing & harvesting	___	___	___	___	___	___	___
zero grazing	___	___	___	___	___	___	___
Total permanent crops	___	___	___	___	___	___	___
of which: vineyards	___	___	___	___	___	___	___
olives	___	___	___	___	___	___	___
(specify) other major_____	___	___	___	___	___	___	___
Ley grassland (under 6 years)							
of which: grazing only	___	___	___	___	___	___	___
grazing & harvesting	___	___	___	___	___	___	___
zero grazing	___	___	___	___	___	___	___
Total green fodder	___	___	___	___	___	___	___
of which: maize	___	___	___	___	___	___	___
(specify) other major_____	___	___	___	___	___	___	___
Total root fodder	___	___	___	___	___	___	___
of which: beet	___	___	___	___	___	___	___
(specify) other major_____	___	___	___	___	___	___	___
Total cereals	___	___	___	___	___	___	___
of which: durum wheat	___	___	___	___	___	___	___
common winter wheat	___	___	___	___	___	___	___
common spring wheat	___	___	___	___	___	___	___
winter barley	___	___	___	___	___	___	___
spring barley	___	___	___	___	___	___	___
(specify) other major_____	___	___	___	___	___	___	___
Grain maize	___	___	___	___	___	___	___
Total root cash crops	___	___	___	___	___	___	___
of which: potatoes	___	___	___	___	___	___	___
sugar beet	___	___	___	___	___	___	___
(specify) other major_____	___	___	___	___	___	___	___
Total oilseed crops	___	___	___	___	___	___	___
of which: sunflower	___	___	___	___	___	___	___
rape	___	___	___	___	___	___	___
(specify) other major_____	___	___	___	___	___	___	___
Other cash crops	___	___	___	___	___	___	___

APPENDIX to CHAPTER 2

SITUATION IN MEMBER STATES REGARDING NITRATE LEVELS IN WATER

BELGIUM

(1) French speaking region

The risk of surface water pollution is due mainly to the excessive and improper use of mineral fertilizers and animal effluent. In some areas it is also due to run off, especially where erosion also occurs. The main causes are nitrates and phosphates.

Risks of deep water pollution are due to the use of soluble mineral fertilizers that separate into ions and are not easily retained by the soil. Nitrates are the main ions produced, along with potassium ions in certain areas. This type of pollution is also caused by excessive use of organic matter that is easily mineralized, such as livestock effluent, sewage sludge and even crops leaving debris with high nitrate content (e.g. leguminous plants).

The table below shows the nitrate concentrations for different soils in water supplies. The figures refer to mixed abstractions (i.e. groundwater and surface water).

App. Table 2.1: Nitrate levels in water from various regions, (1977)

	mg NO_3/litre
Sands of Campine	0.75
Sands of Brussels	13.7
Hesbaye Chalk	12.5
Hainaut Chalk	6.4
Hainaut Carboniferous Strata	0.8
Condroz Carboniferous Chalk	17.1
Average Level in Abstractions	8.1
Minimum	0.0
Maximum	25.5

Current (1988) averages are for 10 mg NO_3/l for groundwater and 13 mg /l for surface water.

The trend over the last 20 years seems to have been a very slow increase (except for entirely isolated incidents, particularly in wells). It is feared that some regions will have a marked tendency toward increased levels due to intensive use of animal effluent. However, use of chemical fertilizers is becoming less excessive.

(2) Flemish region

Surface water pollution is caused mainly by non-farming activities. The majority of groundwater reserves are not yet threatened by intensive livestock farming.

The nitrate content of most deep groundwater boreholes still lies well below the 50 mg NO_3/litre limit. In certain areas, i.e. to the East and South of the Flemish Region, a rapid rise in the nitrate content of deep groundwater boreholes has been noted. Excessive nitrate levels have rendered water from a large percentage of private and shallow groundwater wells no longer fit for consumption.

In certain geological formations, the nitrate content remains low and stable, in other formations the nitrate content is high and rising.

DENMARK

Over 90% of Danish agricultural land is used in the production of rotating crops, the majority being annual crops. Of these, a good 60% are sown in the spring and about 20% are sown in the autumn with winter seed. This means that a relatively large proportion of agricultural land (approx. 60%) has nothing growing on it in the winter period, which gives rise to possible nitrogen loss in run-off into the aquatic environment. Specialisation and the concentration of animal husbandry on fewer farms create problems for the use of farm manure.

Drainwater - The general level of nitrates in drainwater is about 80 mg NO_3/litre, ranging from 60-80 mg in the better soils of Eastern Denmark to 80-100 mg in the sandy areas in Western Denmark. Systematic surveys of drainwater since 1971 have shown that these levels have been relatively constant for the last 17 years. Older unsystematic sample surveys on clayey soil would suggest that the level in 1923-33 was about 40 mg and in 1942 about 65 mg per litre.

Groundwater (drinking water). Denmark has many small reservoirs. In 1983, 65% of the water supplied by the waterworks had a nitrate content (mg NO_3/l) which was under 5 mg. 25% contained between 5 and 25 mg/litre, 8% were between 25 and 50 mg/litre, and 2% of drinking water contained more than 50 mg NO_3 per litre. Nitrate content is generally highest in the western parts of the country.

It is expected that it will be possible to reduce slightly the levels of nitrate in water seeping down below the level of plant roots (drainwater) by means of the restrictions being introduced to reduce nitrogen run-off. This does not necessarily mean that all the risk of an increase in the nitrate levels found in drinking water can be excluded, especially in western parts of the country where groundwater levels are generally high and the reduction capacity of the soil is low and even exhausted in some areas.

FEDERAL REPUBLIC OF GERMANY

Drinking water from the public water supply comes from the following sources: 63 per cent from groundwater, 32 per cent from surface water, and 5 per cent from filtrates from banks of rivers and streams.

Statistics published by the LAWA in 1986 show that in 1983, about 6 per cent of water collection facilities had a periodic or continual nitrate content of more than 50 mg NO_3/litre. The proportion of drinking water exceeding 50 mg NO_3/litre in 1983 is given in the following table.

App. Table 2.2: Proportion of drinking water in Bundesländer exceeding 50 mg NO_3/litre, 1983

	per cent
Schleswig-Holstein	1.0
Niedersachsen	0.3
Nordrhein-Westfalen	8.4
Hessen	0.6
Rheinland-Pfalz	7.0
Baden-Württemberg	3.5
Bayern	3.9
Saarland	1.4
Berlin, Hamburg, Bremen	0

For untreated water from individual wells and groundwater, there is a tendency towards a further increase in nitrate levels. By closing polluted wells (with more than 50 mg NO_3/litre) and opening up deeper ones, water management is trying to comply with legal requirements, so that the amount of contaminated drinking water can be reduced in the future.

Expressed national concern is not so much that some water does exceed the 50 mg/litre level specified in the Drinking Water Authority regulation at both federal and Länder level, but rather the likely upward trends in nitrate levels.

GREECE

Intensive farming has shown a dramatic expansion during the last 20 year-period in Greece. About half a million tonnes of chemical fertilizer nutrients are used yearly as fertilizers while twenty million tonnes of livestock and poultry wastes are produced at the same time on animal farms. Although there is a lack of systematic control and monitoring, it is expected that a great risk of water pollution exists due to either over-fertilization - not uncommon in Greece - or to uncontrolled disposal of animal and poultry wastes. This is increasingly so in some areas which have high density of farms. At present, there are no systematic measurements of nitrate levels at a national or local level.

Concern for water quality - not only from nitrogen - from wastes disposal has resulted in recent regulations and measures to prevent this pollution. As yet, there is no pressure for preventing pollution from fertilizing practices on farms although in some cases, the presence of eutrophication has been detected.

The continuous expansion of intensive farming in Greece will inevitably result in higher-levels of water pollution which in turn, will be expected to result in the appearance of stricter regulations in the near future.

SPAIN

Nitrate content in water varies greatly according to the water supply area and period of the year. Area variations are closely linked to the level of intensive farming. The pattern of farming becomes more intensive from west to east and from north to south. Variations also depend on rainfall levels in each area. Seasonal variations are due to different stages of development for crops, (stages vary greatly from region to region depending on when crops ripen) and to vast seasonal differences in rainfall amounts and subsequent river volumes. Nitrates frequently reach levels far higher than acceptable for drinking water (25-50 mg NO_3/l) in surface water supplies.

The quality of surface water is slowly deteriorating in Spain, according to the graphs in the General Quality Index. This is due mostly to the shortage of flowing water.

The situation with regard to surface water may be summarised as follows:
1. Water quality on the Atlantic slope is acceptable, except in low water periods and in areas with large amounts of sludge e.g. Madrid, Valladolid and Burgos.
2. The Guadalquivir basin is more contaminated due to irrigation, food processing industries and damming.
3. There are problems in the Mediterranean slope due to low amounts of flowing water, quality and salinity due to intensive farming.
4. Reservoirs suffer from severe euthrophication caused by urban and livestock effluents.

In relation to groundwater supplies the current situation is that:
1. Coastal water supplies off the Mediterranean seaboard and the Atlantic in Cádiz, Huelva and the Canary Islands are infiltrated with sea water because of excess abstraction. Saline levels exceed established limits for human consumption and irrigation. This is the case in Tarragona, Castellón, Majorca, Ibiza and Grand Canary Island. Salt water infiltrates supplies in Alicante, Ciudad Real and Valladolid.

2. Nitrates are spreading to underground water in ever wider areas due to the excessive use of nitrogen fertilisers. This occurs mainly in Valencia, Ciudad Real, Seville, Barcelona and Murcia. The situation worsens when water is drawn from contaminated supply to irrigate the surrounding area.

FRANCE

The most widespread water pollution problem in France is caused by farming and in particular, intensive livestock farming which has led to increased nitrate levels. This increase is also due to urban and industial activities which affect mainly surface water. Entire regions are either little affected or not at all, with localised problem areas. In those regions where nitrate levels are already very high, the levels are continuing to climb and in some areas quite rapidly.

IRELAND

Nitrogen fertilizer use in Ireland is among the lowest in the European Community with an average of 65 kg N/ha/year over the utilised agricultural area (UAA) of 5.67 m ha. This low fertilizer N use, coupled with a high average rainfall of 1,150 mm evenly distributed throughout the year, over 90% grassland and little sandy soil contributes to relatively low nitrate levels in water.

The risk of nitrate pollution occurs in small areas with a high proportion of tillage land or intensive dairying with high N use. Private wells contaminated by leaching from farmyards or septic tanks can have high nitrate levels. All public water supplies are below the EEC Maximum Admissable Concentration (MAC) of 50 mg NO_3 (11.3 mg NO_3- N) per litre and with a few exceptions water is generally below the EEC guideline of half this level.

In Ireland about 75% of water for domestic use is taken from surface water. Almost all this water is below the guideline limit of 5.65 mg NO_3-N/litre. The remaining 25% is from groundwater and most of this is also below the guideline limit. There is one borehole in Co Laois, near Carlow town, with 11.3 mg NO_3-N/litre but this supply is no longer in use. There is a well at Balnakill in Co Laois with 11 mg NO_3-N/litre and this is mixed with another well to reduce the concentration. There are three boreholes and two springs near Bagnalstown, Co Carlow with 5.5 to 11.0 mg

NO_3-N/litre and this nitrate concentration is reduced by mixing. A number of boreholes near Kilworth, Co Cork have nitrate levels near the MAC.

It is likely that current levels of fertilizer N use will be maintained but not increased in the immediate future. In some areas, nitrate in groundwater may increase as some of the nitrogen applied in previous years reaches the groundwater.

ITALY

The submitted paper to the CEPFAR conference referred in passing to the province of Mantua as being a designated nitrate risk zone because of its high soil fertility and high livestock intensity. Reference as also made to problems in the province of Emilia associated with the use in agriculture of liquid manures from intensive breeding units, in particular from piggeries.

LUXEMBOURG

Water Supply

In 1985, the average water consumption per day was 117,800 m^3. Two-thirds of this quantity comes from springs and wells, most of which are situated on the geological formation known as the Gres de Luxembourg, which covers the central part of the country. The other third is supplied by SEBES (Syndicat des Eaux du Barrage d'Esch-sur-Sure = the Esch-sur-Sure Dam and Water Authority), whose water treatment plants at the Lac de la Haute-Sure opened in 1969. As only a minority of the districts have sufficient water reserves on their own territory, the drinking water is, for the most part, supplied by the inter-district water authorities.

Nitrate Pollution

On 1 July 1985, 22 localities distributed water with a nitrate content of between 51 and 100 mg NO_3/litre in their territory. These localities were spread over 13 districts, and 7,838 inhabitants (2.1% of the population) received this supply of water. The districts concerned were warned by the Environment Administration and were asked to inform the population of the situation.

Twelve of the 22 localities were able to take immediate action (within days or weeks), by mixing water from various sources in order to reduce the nitrate content of the water supplied. Thus, it was possible to reduce the number of persons receiving water the nitrate nitrate content of which was more than 50 mg NO_3/litre to about 2,800 inhabitants (about 0.8% of the population).

On average, water from the Gres de Luxembourg has a nitrate content of 30 mg/litre. Water supplied by SEBES had a higher nitrate content in 1985 (15 mg/litre) than in 1977 (8 mg/litre). Increasing phosphorous levels have caused eutropication in some lakes.

NETHERLANDS

Recent years have seen an increase in the nitrate content of borehole groundwater from a number of drinking water supply sources. In 1984 two supply sources reached the Maximum Admissible Concentration (MAC) of 50 mg NO_3 (11.3 mg NO_3-N) per litre. Investigations of groundwater quality at local and regional levels have drawn attention to the way in which groundwater can be polluted by nitrates. As well as causing problems in the supply of drinking water, nitrate pollution of the groundwater also contributes to the contamination of nature preservation areas.

A major cause of these problems is farming. This is attributed to the increasing use of artificial fertilizers and the use of animal manure leading to increased leaching of nitrates into the groundwater below farmed land. The concentration of intensive livestock farming on sandy soils, the increased use of artificial fertilizers on grasslands, and the sensitivity of sandy soils to leaching, make the nitrate problem primarily the problem of those regions with sandy soils.

Since 1981 the average nitrate concentration in the groundwater under cultivated land (measured to 10 m below ground) has increased from around 80 to 100 mg NO_3/litre, whilst the average nitrate concentration under grasslands has doubled to approx 20 mg NO_3/litre. Locally, these concentrations can be several tens of milligrams higher. Moreover, nitrates are still finding their way into the groundwater and into drinking water boreholes. It is expected that 25% of all drinking water sources will, in the future, be faced with nitrate problems. In the absence of policy changes, nitrate leaching will, in the future, exceed the MAC limit under approx. 60% of all grassland and 100% of cropland in the sandy area.

PORTUGAL

In general, the nitrate content of surface water is low, and never exceeds 30 mg NO_3/litre in the National Water Quality Network. The highest nitrate contents range from 11 to 21 mg NO_3/litre on major surface water sources such as the Ave, Loures, Azambuija, Mije and Tego rivers.

Little information is available on groundwater as the Pollution Control Administration does not monitor or control nitrate levels. However, data on a section of the Faro Plain under cultivation show high quantities of nitrates. The following are the results of recent (1987) analysis of water samples taken from holes in the ground and pump sites:

    To 40 meters deep: 100-400 mg NO_3/litre
    Over 40 meters deep: 25-380 mg NO_3/litre

It is feared that this situation could become worse, in that Portugal's accession to the EC is expected to lead to increased intensive farming. However, the adoption of the Framework Law on the Environment and impending legislation on water quality should encourage farmers' willingness and motivation to help decrease the risk of pollution due to intensive farming.

UNITED KINGDOM

Currently in the UK all drinking water in public supply has less than 100 mg NO_3/litre but about 2% does not comply with the required EC Maximum Admissable Concentration (MAC). Table 1 shows the recent situation in the 9 Water Authorities of England with regard to the incidence of nitrate levels in water in excess of 50 mg NO_3/litre.

App. Table 2.3: Water sources with nitrate levels exceeding 50 mg NO_3/litre for all or part of the year

WATER AUTHORITY AREA	Surface water		Groundwater		All supplies	
	1985	1986	1985	1986	1985	1986
Anglian	23	22	51	63	30	33
North West	0	0	3	3	0	0
Northumbrian	0	0	0	0	0	0
Severn Trent	5	6	41	48	35	38
Southern	1	0	8	8	1	0
South West	0	0	2	2	0	0
Thames	6	6	12	12	7	7
Wessex	0	0	2	2	1	1
Yorkshire	0	0	13	14	4	3
TOTAL	35	34	132	152	78	82

High nitrate levels occur in the Anglian and Severn Trent areas. Scotland, Wales and Northern Ireland have no water supplies requiring derogations for exceeding the EC MAC. Theoretical models have indicated that nitrate levels in parts of England will continue to rise but, in two aquifers with a short response time, the trend has changed from a rise up to 1981 to a stabilization or slight fall since then. If, as has been suggested, this welcome change is the result of earlier planting of winter cereals and reduced use of nitrogen in autumn and winter, it seems likely that the effect may be widespread.

Animal husbandry in the UK does not cause notable nitrate pollution. Reported incidents of organic pollution of water by animal excreta, silage effluent and dairy wash water have risen over the last few years but this has not been reflected in a parallel change in water quality. About 1% of farmers each year are now involved in such incidents and some 80% of these are caused by dairy farmers.

APPENDIX to CHAPTER 3

Appendix Table 3.1: N production by grazing animals relative to Utilised Agricultural Area

EC Member States and Regions	Total UAA		UAA excluding cereals	
	all animals kg/ha	excluding horses & goats kg/ha	all animals kg/ha	excluding horses & goats kg/ha
BELGIQUE-BELGIE	138	137	184	182
VLAAMS GEWEST	163	160	203	200
REGION WALLONNE	119	118	167	165
BRUXELLES-BRUSSEL	71	71	88	88
ANTWERPEN	253	250	262	258
BRABANT	87	85	149	145
HAINAUT	112	111	166	165
LIEGE	137	136	179	178
LIMBURG	138	136	177	174
LUXEMBOURG	153	152	179	178
NAMUR	98	97	147	145
OOST-VLAANDEREN	181	177	222	217
WEST-VLAANDEREN	152	150	189	187
DANMARK	54	53	124	122
HOVEDSTADSREGIONEN	22	19	60	53
OST FOR STOREBAELT,EX.HOVEDST.	23	22	64	62
VEST FOR STOREBAELT	63	62	137	135
BR DEUTSCHLAND	NA	76	NA	128
SCHLESWIG-HOLSTEIN	NA	85	NA	129
HAMBURG	NA	NA	NA	NA
NIEDERSACHSEN	NA	71	NA	120
BREMEN	NA	NA	NA	NA
NORDRHEIN-WESTFALEN	NA	72	NA	137
HESSEN	NA	65	NA	129
RHEINLAND-PFALZ	NA	52	NA	96
BADEN-WUERTTEMBERG	NA	70	NA	113
BAYERN	NA	91	NA	145
SAARLAND	NA	64	NA	118
BERLIN (WEST)	NA	43	NA	75
ELLAS	NA	32	NA	43
ESPANA	23	20	31	28
NOROESTE	100	94	120	112
GALICIA	96	91	127	120
ASTURIAS	88	80	89	82
CANTABRIA	151	139	154	141

EC Member States and Regions	Total UAA		UAA excluding cereals	
	all animals kg/ha	excluding horses & goats kg/ha	all animals kg/ha	excluding horses & goats kg/ha
NORESTE	21	19	30	28
PAIS VASCO	73	67	93	84
NAVARRA	24	21	33	31
RIOJA	23	21	32	29
ARAGON	15	15	23	22
MADRID	24	22	32	29
CENTRO	19	17	28	25
CASTILLA - LEON	22	21	37	35
CASTILLA - LA MANCHA	12	11	18	15
EXTREMADURA	24	22	29	26
ESTE	21	19	26	24
CATALUNA	26	25	37	35
COMUNIDAD VALENCIANA	10	9	11	9
BALEARES	33	31	41	38
SUR	14	12	19	15
ANDALUCIA	15	12	19	15
MURCIA	12	10	15	13
CEUTA Y MELILLA	NA	NA	NA	NA
CANARIAS	22	9	23	10
FRANCE	52	51	75	73
ILE DE FRANCE	6	6	26	23
BASSIN PARISIEN	43	42	76	74
CHAMPAGNE-ARDENNE	30	29	57	56
PICARDIE	35	34	76	75
HAUTE-NORMANDIE	62	61	99	97
CENTRE	22	21	50	47
BASSE-NORMANDIE	89	87	104	102
BOURGOGNE	47	46	74	72
NORD - PAS-DE-CALAIS	59	59	105	104
EST	58	56	81	80
LORRAINE	58	58	86	84
ALSACE	50	49	95	92
FRANCHE-COMTE	59	58	72	71
OUEST	78	77	105	103
PAYS DE LA LOIRE	85	84	108	106
BRETAGNE	98	97	128	127
POITOU-CHARENTES	49	46	74	70
SUD-OUEST	54	53	75	73
AQUITAINE	48	47	72	70
MIDI-PYRENEES	47	46	67	66
LIMOUSIN	85	84	95	94

EC Member States and Regions	Total UAA		UAA excluding cereals	
	all animals kg/ha	excluding horses & goats kg/ha	all animals kg/ha	excluding horses & goats kg/ha
CENTRE-EST	55	53	67	65
RHONE-ALPES	45	43	57	54
AUVERGNE	66	65	78	77
MEDITERRANEE	17	15	18	17
LANGUEDOC-ROUSSILLON	15	14	17	15
PROVENCE-ALPES-COTE D'AZUR	18	17	20	19
CORSE	18	16	18	16
IRELAND	NA	67	NA	72
ITALIA	42	39	57	53
NORD OVEST	58	56	80	77
PIEMONTE	64	62	95	92
VALLE D'AOSTA	20	20	20	20
LIGURIA	21	18	22	18
LOMBARDIA	110	107	170	165
NORD EST	59	58	81	80
TRENTINO-ALTO ADIGE	37	35	37	36
VENETO	74	72	117	115
FRIULI-VENEZIA GIULIA	43	42	62	61
EMILIA-ROMAGNA	51	50	72	70
CENTRO	25	23	36	34
TOSCANA	24	22	33	30
UMBRIA	26	24	39	36
MARCHE	24	23	43	41
LAZIO	39	35	51	46
CAMPANIA	38	34	50	45
ABRUZZI-MOLISE	31	27	44	37
ABRUZZI	34	29	44	38
MOLISE	25	21	42	36
SUD	22	18	32	26
PUGLIA	15	13	22	19
BASILICATA	25	18	44	32
CALABRIA	33	27	40	33
SICILIA	21	18	28	24
SARDEGNA	48	44	52	48
LUXEMBOURG (GRAND-DUCHE)	NA	107	NA	147

EC Member States and Regions	Total UAA		UAA excluding cereals	
	all animals kg/ha	excluding horses & goats kg/ha	all animals kg/ha	excluding horses & goats kg/ha
NEDERLAND	166	163	183	179
NOORD-NEDERLAND	145	142	161	159
GRONINGEN	89	87	119	117
FRIESLAND	204	202	209	206
DRENTHE	122	120	129	126
OOST-NEDERLAND	209	206	222	218
OVERIJSSEL	205	203	211	209
GELDERLAND	212	208	230	226
WEST-NEDERLAND	125	122	144	142
UTRECHT	242	238	243	239
NOORD-HOLLAND	132	130	145	143
ZUID-HOLLAND	140	137	158	155
ZEELAND	33	32	46	45
ZUID-NEDERLAND	189	184	202	197
NOORD-BRABANT	210	206	222	217
LIMBURG	135	130	150	145
PORTUGAL	NA	NA	NA	NA
UNITED KINGDOM[1]	NA	58	NA	74
NORTH	NA	76	NA	87
YORKSHIRE AND HUMBERSIDE	NA	53	NA	83
EAST MIDLANDS	NA	493	NA	82
EAST ANGLIA	NA	18	NA	43
SOUTH EAST	NA	45	NA	86
SOUTH WEST	NA	92	NA	119
WEST MIDLANDS	NA	90	NA	129
NORTH WEST	NA	107	NA	124
WALES	NA	99	NA	103
SCOTLAND	NA	34	NA	NA
NORTHERN IRELAND	NA	94	NA	NA
Maximum N in the regions listed	253	250	262	258

[1] UAA for regions of the UK is based on 1982 data.
Source: Eurostat (1988) Regions - Statistical Yearbook

Appendix Table 3.2: N production by grazing animals and pigs relative to Utilised Agricultural Area

EC Member States and Regions	All kg/ha	Excluding horses & goats kg/ha
BELGIQUE-BELGIE	182	180
VLAAMS GEWEST	254	251
REGION WALLONNE	124	123
BRUXELLES-BRUSSEL	71	71
ANTWERPEN	333	330
BRABANT	105	103
HAINAUT	117	116
LIEGE	146	146
LIMBURG	204	202
LUXEMBOURG	154	153
NAMUR	100	99
OOST-VLAANDEREN	261	257
WEST-VLAANDEREN	294	292
DANMARK	89	88
HOVEDSTADSREGIONEN	43	41
OST FOR STOREBAELT,EX.HOVEDST.	55	54
VEST FOR STOREBAELT	99	98
BR DEUTSCHLAND	NA	98
SCHLESWIG-HOLSTEIN	NA	103
HAMBURG	NA	NA
NIEDERSACHSEN	NA	101
BREMEN	NA	NA
NORDRHEIN-WESTFALEN	NA	115
HESSEN	NA	83
RHEINLAND-PFALZ	NA	62
BADEN-WUERTTEMBERG	NA	87
BAYERN	NA	104
SAARLAND	NA	72
BERLIN (WEST)	NA	66
ELLAS	NA	34
ESPANA	28	25
NOROESTE	111	105
GALICIA	112	106
ASTURIAS	91	84
CANTABRIA	153	141
NORESTE	26	25
PAIS VASCO	77	71
NAVARRA	30	28
RIOJA	28	26
ARAGON	21	20
MADRID	26	24

EC Member States and Regions	All kg/ha	Excluding horses & goats kg/ha
CENTRO	21	19
CASTILLA - LEON	25	24
CASTILLA - LA MANCHA	14	12
EXTREMADURA	26	23
ESTE	39	38
CATALUNA	55	53
COMUNIDAD VALENCIANA	18	16
BALEARES	39	37
SUR	18	15
ANDALUCIA	17	14
MURCIA	26	25
CEUTA Y MELILLA	NA	NA
CANARIAS	26	13
FRANCE	56	55
ILE DE FRANCE	7	6
BASSIN PARISIEN	45	44
CHAMPAGNE-ARDENNE	31	30
PICARDIE	36	36
HAUTE-NORMANDIE	64	63
CENTRE	23	22
BASSE-NORMANDIE	91	89
BOURGOGNE	48	48
NORD - PAS-DE-CALAIS	68	67
EST	59	58
LORRAINE	59	59
ALSACE	54	53
FRANCHE-COMTE	61	60
OUEST	89	88
PAYS DE LA LOIRE	89	88
BRETAGNE	127	126
POITOU-CHARENTES	51	48
SUD-OUEST	57	56
AQUITAINE	52	50
MIDI-PYRENEES	50	49
LIMOUSIN	87	86
CENTRE-EST	57	55
RHONE-ALPES	48	46
AUVERGNE	68	67
MEDITERRANEE	18	16
LANGUEDOC-ROUSSILLON	16	15
PROVENCE-ALPES-COTE D'AZUR	20	18
CORSE	19	16
IRELAND	NA	69

EC Member States and Regions	All kg/ha	Excluding horses & goats kg/ha
ITALIA	48	45
NORD OVEST	63	61
PIEMONTE	70	69
VALLE D'AOSTA	21	20
LIGURIA	21	18
LOMBARDIA	137	134
NORD EST	65	64
TRENTINO-ALTO ADIGE	38	36
VENETO	81	80
FRIULI-VENEZIA GIULIA	50	48
EMILIA-ROMAGNA	69	68
CENTRO	31	29
TOSCANA	29	27
UMBRIA	35	33
MARCHE	30	29
LAZIO	41	37
CAMPANIA	41	37
ABRUZZI-MOLISE	33	29
ABRUZZI	36	31
MOLISE	28	24
SUD	23	19
PUGLIA	16	14
BASILICATA	28	21
CALABRIA	35	30
SICILIA	22	19
SARDEGNA	50	46
LUXEMBOURG (GRAND-DUCHE)	NA	113
NEDERLAND	234	231
NOORD-NEDERLAND	154	152
GRONINGEN	96	94
FRIESLAND	210	208
DRENTHE	138	135
OOST-NEDERLAND	300	296
OVERIJSSEL	281	278
GELDERLAND	314	310
WEST-NEDERLAND	143	141
UTRECHT	311	307
NOORD-HOLLAND	137	134
ZUID-HOLLAND	159	156
ZEELAND	40	38
ZUID-NEDERLAND	374	369
NOORD-BRABANT	408	403
LIMBURG	290	285
PORTUGAL	NA	NA

EC Member States and Regions	All kg/ha	Excluding horses & goats kg/ha
UNITED KINGDOM[1]	NA	63
NORTH	NA	77
YORKSHIRE AND HUMBERSIDE	NA	68
EAST MIDLANDS	NA	49
EAST ANGLIA	NA	33
SOUTH EAST	NA	52
SOUTH WEST	NA	97
WEST MIDLANDS	NA	95
NORTH WEST	NA	116
WALES	NA	100
SCOTLAND	NA	35
NORTHERN IRELAND	NA	101
Maximum N for the regions listed	408	403

1 UAA for regions of the UK is based on 1982 data.
Source: Eorostat (1988) Regions - Statistical Yearbook

APPENDIX to CHAPTER 4
FERTILIZER USE PRACTICES

BELGIUM

(1) French speaking region

Chemical Fertilizers

There are three distinct types of practice to be distinguished when referring to fertilizer use:

1. **Habitual Use:**

    This happens especially with P and K. For a long time, these farmers used set amounts of fertilizers, greater than crop output requirements on a regular basis. This also happens with nitrogen fertilizers, in areas where farming is not so intensive and in extensive or semi-extensive grazing areas.

2. **Semi-Rational Use:**

    This type of use adapts amounts according to ideas popular at the time. For example, in some areas with soils rich enough or too rich in P and/or K, officials have recommended no fertilizer use. Likewise, the farming advisors' campaign to decrease amounts of nitrogen fertilizer applied on land is beginning to take effect.

3. **Rational Use:**

    This is based on concrete data (soil analysis, awareness of reference standards, crop rotation...), estimates of soil productivity, knowledge of fertilizing substances and consultation of various sorts on fertilizer advice (such as nitrogen manure in the Gembloux Cereals Reference Register) offered by objective specialists. Rational use is now growing, due to the higher training level of farmers, the decrease in number of farmers with consequent greater ease in informing them and perhaps the smaller profit margins for most farm enterprises in recent years.

The table below shows trends in chemical fertilizer consumption for the region.

App. Table 4.1: Chemical fertilizer consumption 1970-1985,

Element	1970	1975	1980	1985
		kg/hectare of farmland		
Nitrogen (N)	102.5	108.1	127.1	129.1
Phosphorus (P_2O_5)	86.7	71.1	67.0	60.2
Potassium (K_2O)	109.2	88.0	98.6	89.3

Animal Manures

A distinction should be made between two types of intensive livestock farming :

1. Intensive farming of various animal species on farms with vast tracts of land, and
2. industrial" or "off soil" intensive farming where each farmer has little or very little farmland; the problem of land scarcity on these farms is further aggravated by the fact that they are concentrated in confined areas (eg port regions, regions with many small farms).

   The slurry spreading problems on the first type of farms are limited to :
   - odour, relations with neighbours.

   - surface run-off or loss through percolation on the site where effluent is produced.
   - manuring dates do not always coincide perfectly with crop requirements but are dictated by slurry production conditions (season) and the size of storage units.

In general, it is believed that a balance has been struck between effluent production and the possibility for its efficient and non-polluting use.

The problems are many and acute on the second type of farm. They include all those listed above but to a greater extent. At times, it is impossible to grow crops because the soil is actually poisoned by a glut of effluent. In general, it is believed that the balance between animal farming and the possibility of using effluent rationally without polluting has been upset. The balance between production (animal and crop) and consumption of energy and nutritional proteins has also been upset, both on the farm and regionally, (commune, region, territory).

(2) Flemish region

Chemical N Fertilizers

In order to protect water supplies, N use is limited to 400 kgs per hectare and banned within a 2 km radius of groundwater collection points during autumn and winter (ie from September 1 to January 31).

Animal Manures

Manure storage capacity runs to 2 - 3 months on existing farm units. Limited storage facilities mean that manure continues to be spread during autumn and winter.

DENMARK

Chemical Fertilizers

Fertilizers are used on the basis of results of research and experimentation. About 2,500 field trials are carried out on farms each year under the supervision of the advisory services of the national agricultural organizations. Some 500 of these are fertilizer trials and about 300 of these involve nitrogen fertilizers. The results are available during January; about 40,000 copies of the report on the trials are printed. This report is the most important handbook available to advisers and farmers on the subject of fertilizer planning and many other aspects of crop production in arable farming.

The advisors are closely involved with farmers' fertilization planning as regards quantities, times and methods of application. The advisers have also prepared 20,000 to 22,000 detailed fertilization programmes each year, representing about 40% of agricultural land. This figure is likely to reach 30,000 to 35,000 in 1987/88 due to the requirement that all farmers, as and from May 1988, should be able to produce a fertilization plan at the request of the authorities.

Most advisers nowadays prepare fertilization programmes with the aid of either a PC programme or a central EDP programme handling data recording and storage. The remainder of the legally required fertilization programmes are prepared manually by the farmers themselves, and a few are prepared with the help of bulk-distribution companies.

Since 1976 the advisory services of the agricultural organizations have issued regional forecasts for nitrogen needs in the following growing season. This forecast is now based on about 800 soil samples taken by "KVADRATNET", Denmark's nitrate survey, and on temperature and precipitation figures for the September-March period.

The target for nitrate application is the economically optimal quantity of nitrogen per field, and to this end, all practical management methods, including N-forecasts, N-min and plant analysis, are used.

Application of phosphorus, potassium and other mineral fertilizers is done on the basis of annual field soil tests.

Synthetic nitrogen consumption has increased up to the early 1980's (see table below) with the national average in the range 135 to 142 k N/ha in the last five years.

App. Table 4.2: Total Use of Fertilizers, Pure Nutrients in Denmark

	1970/71 -74/75	1975/76 -79/80	1983 /84	1984 /85	1985 /86
Commercial fertilizers, 1000 t :					
Nitrogen (N)	318	367	412	398	382
Phosphorus (P)	59	59	52	49	46
Potassium (K)	158	143	130	124	121
Manure, 1000 t :					
Nitrogen (N)	141	157	171	168	172
Phosphorus (P)	52	68	70	68	70
Potassium (K)	150	179	194	192	189
Total consumption, 1000 t :					
Nitrogen (N)	459	524	583	566	554
Phosphorus (P)	111	127	122	117	116
Potassium (K)	308	322	324	316	310
Total consumption kg per ha :					
Nitrogen (N)	156	179	205	198	195
Phosphorus (P)	36	43	43	41	40
Potassium (K)	102	110	114	111	110
Commercial fertilizers as percentage of total consumption :					
Nitrogen (N)	70	70	71	70	69
Phosphorus (P)	55	46	43	42	40
Potassium (K)	53	44	40	39	39

Animal Manures

Developments in farm management have been characterised, for a number of years, by increased specialization in farming either with or without livestock and in specialization within livestock farming between dairy, cattle, pig and poultry farming. The numbers of the old type of farms with mixed livestock continues to decrease.

There have also been differing developments in different parts of the country. A large proportion of livestock has moved away from the better soil types on the islands and onto the sandy soil of Western Jutland, where irrigation is also possible.

There is also an increasing problem with pig farming in that it is carried out with bought-in feedstuffs and is thus independent of farm size. This contrasts with cattle farming where animal manure produced on the farm can be disposed on the land and the level of manure production would be generally related to the size of the farm. In Western Denmark it is not possible to buy extra fodder in the form of waste from brewers, sugar factories, neighbours' sugar beet tops, etc. Almost all breweries and sugar factories are situated in Eastern Denmark.

So it is primarily pig farmers who may have difficulties in making sensible use of animal manure on their own property. In some cases poultry farmers face similar problems. There are, however, farms with so big a dairy herd in proportion to their acreage that proper use of animal manure is not feasible, so that surplus manure must be either sold or given free of charge to other properties.

The table above shows that the amount of N from animal manures sources, has increased in line with synthetic fertilizers, such that the latter continue to account for 70 per cent of total N consumed. The amount of N in animal manures is estimated at 170,000 tonnes, which results in an effective N in the field estimated at 50,000 tonnes or 18 kg per ha.

FEDERAL REPUBLIC OF GERMANY

Fertilizer practices take into account of recommendations and technical advice provided by federal and/or regional authorities. These include:
1. Local experience and monitoring the fertilizer needs of growing crops
2. Mineralised nitrogen measurements for early year fertilization
3. Fertilizer modelling using computers
4. "Nitrogen help table" which relates current crop nitrogen requirements to pertinent local factors
5. Plant analysis to determine nitrogen supply to growing crops.

Chemical fertilizers

Chemical nitrogen fertilizer consumption has stagnated since 1980

App. Table 4.3: Chemical fertilizer consumption in Germany, 1980-87

Year	Million tonnes N	Kg N/ha agric. land
1980	1.47	120.0
1981	1.55	126.6
1982	1.32	108.5
1983	1.46	120.7
1984	1.38	114.1
1985	1.45	120.5
1986	1.52	126.1
1987	1.58	131.5
Average 1980-87	1.47	121.0

Source: Federal Statistical Office in Wiesbaden, Fachersie 4, December 1987

Actual consumption of fertilizers in individual regions is difficult to ascertain due to the high levels of trade between Länder. It is known, however, that Schleswig-Holstein, which has a relatively high nitrogen consumption, has few problems with high nitrate content in groundwater, while some Länder with low nitrogen consumption (e.g. Rheinland-Pfalz, Baden-Württemberg) have more serious nitrate problems. It is considered that there is not necessarily a direct connection between the amount of chemical nitrogen fertilization and nitrate content in groundwater (hydrology, cultivation, etc are more decisive in this context).

Animal manures

Based on livestock population statistics, an annual average of 75 kg per hectare of total nitrogen has been calculated. This does not take into account gaseous NH_3 loss during spreading. It should be noted however that livestock farming is not evenly distributed throughout the country.

Other sources of nitrogen

The amount of nitrogen supplied from the atmosphere is between 15 and 25 kg/ha; biologically fixed nitrogen quantities are of a similar order of magnitude. In contrast, there is about 20 to 40 kg/ha of gaseous wastes from nitrogen due to denitrification and NH_3 volatilisation, as well as unavoidable nitrogen loss from leaching. These sources of nitrogen should be compared within an overall balance of nitrogen to the amount of nitrogen taken out of soil by crops.

These overall nitrogen calculations can only give information on an excess or deficit of nitrogen of a large area and say nothing about the danger of nitrate displacement from a particular area. Therefore, they are not really usable in trying to reduce or control regional nitrate pollution.

In order to maintain soil fertility, the amount of nitrogen fertilizer put on a crop should be calculated at about 20 to 30 kg/ha above the nitrogen taken out of the soil from the crop (based on cereals, beets, and tubers).

GREECE

Chemical Fertilizers

The use of fertilizers in Greece has increased in line with the increase in intensive farming. Organic fertilizer use has remained at a very low level or has even been, in recent years, abandoned. Despite their relatively high cost, there is a tendency for chemical fertilizers to be applied increasingly by farmers against the advice of officials. This could be attributed to the fact that cultivated soils each year become poorer in terms of fertility due to the lack of organic fertilizer use. As a result, a greater risk of water pollution is expected to occur due to excessive run off. Chemical fertilizers are commonly spread either directly or following their dilution with irrigation water. The evolution of chemical fertilizer use at a national level is shown in the table below.

App. Table 4.4: Evolution of Chemical fertilizer use in Greece

Year	Fertilizers (nutrient) use, kg/ha of Farmland		
	N	P_2O_5	K_2O
1974/75	71,0	42,4	6,9
1975/76	77,0	45,2	7,8
1976/77	82,2	49,5	10,6
1977/78	87,8	50,3	10,8
1978/79	93,7	53,5	12,0
1979/80	95,3	48,3	11,4
1980/81	99,5	44,6	10,2
1981/82	105,3	46,7	11,6
1982/83	115,2	49,8	13,3
1983/84	113,3	56,1	13,1
1984/85	123,1	50,9	15,8
1985/86	120,3	49,9	17,0

Animal Manures

The increased need for livestock and poultry products in Greece has led to the installation of a large number of intensive livestock enterprises. Intensive farming in Greece today is concerned mainly with poultry and pig production. Dairy and beef farms are run at a lower level of intensification while sheep and goats remain at an extensive stage due to local conditions.

Livestock and poultry wastes are produced in the form of solid wastes (poultry and part of dairy and beef wastes) and in the form of liquid wastes (pig and the other part of dairy and beef wastes).

Because of the pattern of intensive farming in Greece (poultry and pig farms) both solid and liquid wastes are capable of polluting water. But the generally drier climatic conditions and the larger quantities produced make liquid wastes from pig farms the most probable water pollutant. Direct disposal in certain humid areas, with intensive animal farms, gives rise to the danger of pollution from both solid and liquid wastes.

SPAIN

Chemical fertilizers

Spanish farmers use fertilizers with little regard to technical information. Only in horticultural areas in Levante and the South are there highly sophisticated farms regulating fertilizer amounts, but without any official monitoring of environmental effects.

The following table shows the trend in fertilizer consumption since 1978.

App. Table 4.5: Consumption of fertilizer in Spain, 1978-86

Year	Farm area '000 ha	N '000 t	N kg/ha	P_2O_5 '000 t	P_2O_5 Kg/ha	K_2O '000 t	K_2O kg/ha
1978	16 955	793	46.8	434	25.6	273	16.1
1979	16 794	903	53.8	456	27.2	283	16.8
1980	17 057	985	57.7	473	27.8	295	17.2
1981	16 965	806	47.5	420	24.8	254	15.0
1982	17 222	884	51.3	405	23.5	255	14.8
1983	17 216	688	39.9	366	21.3	225	13.1
1984	17 379	870	50.1	429	24.7	277	15.9
1985	17 300	942	54.4	463	26.8	304	17.6
1986	17 753	890	50.1	426	24.0	286	16.1

Animal manures

Disposal of animal manures varies widely from region to region.

Excreta are distributed as follows:
- 20 per cent is used as manure after fermentation
- 20 per cent is used on grazing soil
- 10-15 per cent is purified through various metals. A small portion is treated in biogas facilities.
- Reminder: directly spread on ground.

On mixed farms - crop and livestock farming - or livestock farms with available land, the usual practice is to spread the excreta as slurry on the ground using pump tanks.

On land-independent farms, practice varies depending on the municipal authority's capacity to monitor it. In districts with sufficient administrative capacity, regulations which deem manures in certain circumstances as dangerous have led to the movement of these manures to more rural areas where it is a normal practice to manure with little or no official supervision or control.

Where farmers have not transported manure farther away, the usual practices are as follows:
- leasing land to spread excreta
- reaching agreements and conventions with nearby farmers for the same purpose
- occasionally the local authority can approve or facilitate spreading of manure on gravel pits and other holes in the ground left by abstraction equipment
- making special ponds for spreading excreta, which is left to dry and removed afterward
- a small portion is used in biogas production

These situations almost exclusively relate to pig excreta. Nationally, about seventy per cent of all pigs are farmed intensively. Although some bird farms use the excreta to spread slurry on the soil, the tendency is to re-use excreta in ruminant feed.

Regionally, the most affected areas are the Cantabrian Mountains and Navarre, Levante and Catalonia due to the high concentration of intensive livestock farming.

FRANCE

Chemical Fertilizer

The use of Nitrogen has been increasing (see table below) but the rate of increase has slowed in the 1980's. The average level of N use in 1985 was 81 kg/ha.

App. Table 4.6 : Chemical nitrogen fertilizer use in France

Year	1963	1970	1975	1980	1985
N (kg/ha)	24	42	53	73	81

Animal Manures

The notion of intensive farming must be clarified in the case of France. A distinction should be made between (i) production which is intensive but where the animal manure can be disposed of on the land, (ii) situations where the amount of manure produced is very high, sometimes too high for the amount of land that is available to receive it. The latter areas are located mainly in Brittany. Farmers there must find land surfaces for manure spreading outside their farms, sometimes rather far away from them. In France, as elsewhere, farmers who raise livestock generally combine the use of livestock effluent, which is very rich in fertilizers, with the use of artificial fertilizers. Compost and sludge from factories and urban areas are also used in some places. If farmers do not raise animals, they rarely use the manure from other farmers, except in certain areas.

Nitrogen from animal manure increased up to the mid 1970's but has been stable since then, at 46 kg per hectare.

IRELAND

Chemical Fertilizers

The pattern of fertilizer use in the Republic of Ireland over the past 30 years is summarized in the table below. Fertilizer use followed a generally upward trend until 1973/74. After that the use of nitrogen increased rapidly as a result of greater intensification on dairy farms and a more widespread production of silage for winter feed. This has continued up until the present time but at a slower rate since 1984. The use of phosphate and potash has tended to even out and any fluctuations which have occurred have been caused by rises and falls in prices. Their usage at present is not much different from what it was in 1973/74.

App Table 4.7: Fertilizer Sales in the Republic of Ireland, nutrient tonnes

Trading Year	N	P	K	Total
1972/73	131,775	94,774	155,346	381,865
1973/74	130,208	84,306	151,025	365,539
1974/75	133,044	50,529	93,111	276,684
1975/76	152,739	58,747	120,206	331,692
1976/77	167,461	65,186	141,638	374,285
1977/78	230,214	76,347	170,397	476,958
1978/79	263,603	80,335	183,836	527,774
1979/80	247,535	67,965	157,010	472,510
1980/81	275,058	63,134	150,349	488,541
1981/82	275,186	61,819	147,949	484,954
1982/83	295,985	63,391	153,216	512,592
1983/84	331,440	66,203	161,641	559,284
1984/85	327,709	66,028	163,811	557,548
1985/86	322,747	58,083	144,690	525,520
1986/87	371,656	65,887	165,495	603,038

The average rates of fertilizer use (kg of nutrient/hectare) in 1986/87 were:
- Nitrogen (N)   65
- Phosphorus (P)   12
- Potassium (K)   29

Grassland receives between 0 and 400 kg of nitrogen, between 0 and 40 kg of phosphorus and between 0 and 75 kg potassium per hectare per year. On average, tillage areas receive higher rates of fertilizer than grassland because many dry stock farms use little or no N on grassland. Cereals receive from 0 to 200 kg of nitrogen, 0 to 45 kg of phosphorus and from 0 to 95 kg of potassium per hectare per year and potatoes and root crops range between 60 and 220 kg of nitrogen, 80 and 150 kg of phosphorus and from 50 to 110 kg of potassium per hectare per year.

Animal Manures

Recent developments in agriculture include the intensification of livestock production, associated with enlarged pig and poultry units, the housing of cattle in winter, increased and more widespread silage production. As a result of these developments the disposal of farming wastes such as animal manures and silage liquor are assuming greater importance as sources of pollutants in the aquatic environment.

Between 1975 and 1987 pig numbers increased by 25 percent and silage production by 170 percent. Past experience has shown that pig rearing and silage production have the potential for serious pollution.

LUXEMBOURG

The water pollution linked to agricultural practices differs as between the springs located at the Gres de Luxembourg and the Lac du la Haute-Sure.

The former springs are found in an area where soil is susceptible to leaching to begin with - where the topsoil is very light - and in addition, maize growing represents a substantial part of the crop rotation. In some cases, the easy access to these lands in winter means that the amount of liquid fertilizer spread is several times higher than the national average of 13-14 m^3/hectare/year. Furthermore, as there is little awareness of the problems resulting from lack of storage (storage volumes are estimated to be sufficient for three months manure production) the spreading of fertilizer is often done under unacceptable conditions (eg soil saturated with water, beginning of winter, land covered with snow, etc). It should be noted that the storage capacity of the new liquid fertilizer storage basins is 5 months.

Water pollution in the Lac de la Haute-Sure is a problem of greater complexity. Part of the problem is the large quantity of nutrients used in agriculture, especially nitrates leached from schistous soils in Oesling (in the northern part of the country), phosphates from improper spreading of mineral fertilizers (basic slag), and liquid fertilizer on fields with a steep gradient under climatic conditions when run-off is likely. Moreover, "accidents", such as fertilizer or liquid manure spills, also occur from time to time.

Another aspect is the lake's geographic location. Seventy percent of the basin draining into the Sure is on Belgian territory. The fact that the Sure's water enters Luxembourg without having been treated in any way poses significant problems in the production of drinking water.

NETHERLANDS

Chemical Fertilizers

The consumption of chemical fertilizers by the farming industry is related to the type of farm (eg livestock, crop, horticulture) and the farm methods applied (eg intensive/extensive, farming mix). The table below provides an overview of the consumption of chemical N fertilizers in the sandy regions of the Netherlands in 1979/1980. Since this survey was undertaken, horticulture has expanded by approximately 8,000 ha and approximately 30,000 ha of grassland have been turned over to crops (primarily fodder maize). During the same time total chemical N fertilizer consumption in the Netherlands has increased by approximately 10 million kg N to 495.7 million kg N for a land surface of around 2 million ha.

App. Table 4.8: Estimated Chemical Fertilizer Consumption on Grassland and Croplands in the Sandy Regions of the Netherlands, 1979/1980

	Area (ha)	Kg N/ha	Eastern Area (ha)	Kg N/ha	Central Area (ha)	Southern Kg N/ha
Grassland:	161,892	274	67,906	195	167,176	295
Cropland:	38,507	103	9,868	87	103,982	71
Horticulture:						
Cat. 1*	141	119	125	131	9,198	135
Cat. 2**	235	68	448	69	4,861	69
Glass	26	502	44	528	1,082	616

NOTES: * cat 1. = annual plants
** cat 2. = biennials or perennials
Glass = cultivation under glass

On most grassland, animal manure is also applied in addition to chemical fertilizer, therefore bringing the total N application to more than 400 kg N/ha/year. Once total N application exceeds about 300 kg N/ha/year, the level of nitrate leaching from grassland appears to rise sharply. In addition, grazing also contributes to nitrate leaching. The ultimate level of nitrate pollution of shallow groundwater depends, among other things, on the groundwater level and its denitrification capacity. The table shows a clear distinction in the use of artificial fertilizers between intensive and extensive dairy cattle farming. In the Eastern and Southern sandy regions the application of nitrogen per hectare of grassland is higher than in the central sand area.

The relatively low levels of artificial fertilizer on sandy croplands, compared with other land users, is due to the widespread cultivation of fodder maize, where slurry is used as fertilizer. However, this form of fertilizing also leads to high levels of leaching underneath maize crops. Other forms of cultivation in these areas play only a minor role in nitrate leaching.

The same applies in general to horticulture, because of the small area given over to this form of land use. However, given its high consumption of artificial fertilizer per hectare, horticulture can, at the local level, contribute substantially to nitrate pollution. It is a known fact that between 60 and 600 kg of nitrate/ha/year can be lost in one way or another from horticultural units.

Animal Manures

Since 1950 the livestock population has grown substantially, with a current population of approx. 14 million pigs, 100 million head of poultry and 4 million cattle. In 1986, these animals produced 95 million tonnes of manure, with a nitrogen content of 481 million kg.

The concentration of intensive livestock farming in certain parts of the country has led to the formation of so-called manure surplus areas. The highest manure production levels are found in the sandy regions. Allowing for some of this production to be absorbed by crops (at the environmental protection level), a nitrogen surplus of up to 660 kg N/ha exists in these regions. Animal manure is applied principally to grazing land and fodder maize crops. Elsewhere application depends on the type of crops. In practice, around 50 tonnes of slurry, mostly from cattle, are applied per hectare of grassland per year. Fodder crops receive a full range of manures in varying quantities. Maize boasts a strong resistance to high nitrogen levels. It is by no means unusual to find slurry levels on maize fodder crops well in excess of the necessary provision of minerals. On grassland, concentrations of up to 300 mg NO_3/litre have been measured in the upper groundwater (circa. 1 m below surface level), caused by the application of the recommended level of 400 kg N/ha/year and frequently also due to the fact that when applying artificial fertilizer, insufficient attention is paid to the release of nitrogen from animal manure.

Independently of the type of manure used, leaching losses of between 150 and 400 N/ha/year on maize land have been found. (N.B. On ground with a deep groundwater level, no denitrification and a rainwater surplus of 300 mm/year, a leaching loss of 34 N/ha/year is equal to 50 mg NO_3/litre). In part, this is a result of the high basic leaching level of approx 50 kg N/ha/year on cropland. This high basic leaching is the result of years of heavy application of fertilizers and will tend to decrease (by some unknown proportion), with reduced application levels. The extent to which high concentrations in the surface groundwater lead to high concentrations in deeper groundwater varies from region to region, depending on the geohydrological soil structure, denitrification, groundwater currents, surplus rainfall levels etc.

PORTUGAL

Chemical fertilizers

The usual pattern of use is basic fertilization in the autumn with low nitrogen mixtures. Fertilizers high in nitrogen and phosphorous are generally applied in the spring. Green manure is used to a limited extent.

In 1985, 429 624 tons of nitrogenous fertilizers were used, and in 1986 656 481 tons. Data on the use of fertilizers at regional level are not available.

Animal manures

Little or no information on the availability and use of animal manures is available. There is little intensive livestock farming. Outdoor housing of cattle and pigs is still practiced to a large extent and it is considered that as a result, few pollution problems arise. Animal manures are spread mainly by flooding land or in strips and are used principally in vineyards and pasture.

UNITED KINGDOM

Chemical Fertilizers

Fertilizers are used according to need as indicated by field experiments and by individual farmer's experience. Advice on application rates is available from the industry and, independently, from the Advisory Service of the Ministry of Agriculture. This advice takes into account, in general terms, the nitrate producing ability of the soil.

Since 1969, the Government Advisory Service, the Agricultural Research Council, the Scottish Agricultural Colleges and the industry have collaborated in a detailed survey of fertilizer practice, separately for Scotland and for England and Wales. This permits an assessment of use by crop, by farming systems, by month of application (in recent years) and by fertilizer type.

This long time series provides an invaluable data base for assessing trends and for checking actual usage against recommendations. The average actual usage is close to the recommended level. Although some farmers are above the average, others are below and the net effect in terms of nitrate leaching should reflect this.

The greatest risk of nitrate leaching occurs in autumn when fields become saturated with water. At this time nitrate formed from organic nitrogen compounds in the soil accumulates if it is not being taken up quickly by growing plants. Nitrogen applied under these circumstances adds to the risk of nitrate leaching. Both industry and the advisory service have, in the last ten years, been advising against use of nitrogen fertilizers in the autumn and use has fallen by about half in the 1980's. Farmers have no legal obligation to restrict fertilizer use but are responding to advice on good agricultural practice. There is scope for further reduction but already this development is an important factor in the change from rising nitrate concentration to constant or falling levels. The other change which has made at least as important a contribution is the earlier sowing in autumn of winter cereals.

About 90% of the crop and grass area received nitrogen applications. The average application rate for England and Wales is 163 kg N per hectare but it varies by region in the range 100 to 190 kg per hectare.

Animal Manures

In the UK, pigs and poultry are reared to a considerable degree, in intensive systems. This is generally not the case for cattle. There is no significant nitrate pollution from manures or other organic material. Most manure is spread within the same local district (parish) as the farm unit. In a few areas, pig manure is "exported" to a neighbouring parish. But nowhere is the intensity of pig rearing anything like as great as in countries such as the Netherlands. There are a few areas of cattle farming in the East and Midlands where the highest risk of nitrate pollution exists.

APPENDIX to CHAPTER 6

EFFICIENCY INDICES BASED ON MIX OF MINERAL/ORGANIC NITROGEN

An basis for estimating efficiency indices from the mix of mineral/organic nitrogen in manures is given in the CEC 1978 report[1]. Table A6.1 gives estimates based on the assumption that humus formation from manure organic matter has reached equilibrium, so that the breakdown of old humus is offset by an equal amount of humus build up.

Table A6.1: Efficiency index for animal manure nitrogen under humus equilibrium, with the corresponding values from Table 6.1 in brackets

Animal Manure (% Mineral N)	Arable land		Zero grazed Grassland	
	Spring (%)	Autumn (%)	Spring (%)	Autumn (%)
Liquid manure (94%)	79 (80)	34 (40)	70 (70)	31 (35)
Slurry (50%)	73 (50)	48 (25)	81 (35)	58 (20)
Farmyard manure (10%)	67 (40)	60 (20)	91 (20)	82 (10)

Before this equilibrium is reached, humus build up from organic matter in manure would exceed humus breakdown, so that less nitrogen would be available from manure organic matter sources. Table A6.2 shows efficiency indices based on first year application of animal manures.

Table A6.2: Efficiency index for animal manure nitrogen in the first year of application, with the corresponding values from Table 6.1 in brackets

Animal Manure (% Mineral N)	Arable land		Zero grazed Grassland	
	Spring (%)	Autumn (%)	Spring (%)	Autumn (%)
Liquid manure (94%)	77 (80)	32 (40)	67 (70)	28 (35)
Slurry (50%)	55 (50)	30 (25)	57 (35)	34 (20)
Farmyard manure (10%)	35 (40)	28 (20)	47 (20)	39 (10)

The efficiency indices from Table 6.1 are shown in brackets to facilitate comparison.

---

[1] Commission of the European Communities (1978a) *The Spreading of Animal Excrement on Utilized Agricultural Areas of the Community*, Volume I, Information on Agriculture, NO. 47, pp 16-16.

APPENDIX to CHAPTER 8
POLICIES AND MEASURES TO CONTROL NITROGEN POLLUTION IN MEMBER STATES
BELGIUM

(1) **French speaking** region

Chemical Fertilizers

Currently there are no regulations regarding the use of chemical fertilizers. The approach to protecting water from fertilizer pollution involves:

1. Increase in knowledge of soil composition through analysis and application of results according to soil type and cultivation methods, increase in knowledge of nitrogen cycle in farmed land as well as in non- farming areas (e.g. forests, fallow land, urban or industrial areas).
2. Official advisors' recommendations for better use with regard to:
    - Fertilizer amounts used,
    - Proper fertilizer application, including timing, absorption into tilled layer, avoidance of spreading on frozen soil or in periods of major drainage, etc.

Expectations are firstly that financing of research and dissemination of information on these issues, will be extended for the foreseeable future. Regulations may follow provided they are formulated with due regard to good judgement and cooperation with neighbouring areas.

Animal Manures

There are no regulations on animal manure but regulations and organizational assistance are expected in the medium term to improve production, storage, transfer and use of effluent from intensive livestock farming.

(2) **Flemish region**

The use of nitrogenous fertilizers is limited to 400 kg N per hectare. The animal population is limited to 4 adult cattle equivalents per hectare and

permission is required for the storage of slurry in both underground and above-ground silos within a radius of 2 km of groundwater collection points as well as for the use of raw fodder silos (ground and trench silos).
The spreading of nitrogenous fertilizers (including slurry) is banned from September 1 to January 31 within a 2 km radius of groundwater collection points. There is an import ban on (foreign) slurry.

Newly built animal housing is required to contain 6 month's manure storage capacity. Since February 1987 there is a flexible limitation on the size of indoor livestock units which have no outdoor grazing facilities. Existing units are allowed to expand to a maximum of 1,000 pig and 300 calf stalls (meat calves). Under pressure from drinking water and water purification companies, the flexible building limitation has led to a complete ban on additional indoor units in approximately half the Flemish region. The authorities are hoping that this will allow them to get the manure problem under control. Existing farms are no longer permitted to change to indoor farming. New units may not be set up. Various units have been refused permission to exploit farm buildings for which they have received building permits. Attention is being paid to the rational utilization of slurry, and in particular to setting up manure banks and the advisory services

The Flemish government is preparing an order concerning animal manures. it is proposed that:
1. The Minister for the Environment will be empowered to decide when, how much and how slurry may be applied. Definite proposals for specific limits in these areas have not as yet been drawn up.
2. The Minister will place a manure levy on each farming unit based on the number of animals and the production of manure per hectare.
3. Provision will be made for the extension of one or more manure depots to act as intermediaries between manure surplus and manure deficit areas.

In the longer term, the promotion of 'good farming' is expected to solve most of the pollution problem. Experimental fieldwork (simplified) and advisory services are two major instruments of this policy. Applied scientific research could improve use of minerals by animals and lead to an economically viable form of manure processing It is considered that not enough is known about the inter-relationships between manure, fertilizers, soils, plants, water and water-extraction. An economically and technically viable solution of the nitrate problem must take thorough account of these inter-relationships.

DENMARK

Chemical Fertilizers

At present there are no laws, regulations or controls on the use of nitrogen from chemical fertilizers, except for some areas which are protected as nature reserves, border on special lake areas or are sensitive areas for groundwater collection.

A Ministry of Agriculture Notice (No. 655 of 9 October 1987) introduced a requirement that all farmers must establish fertilization programmes for their land. There is also a requirement that 45% of the farm area must in 1988 have vegetation on it in the autumn months up to October 20. This proportion is to increase to 55% in 1989 and to 65% in 1990.

Farms with large cattle herds, large areas under grass, or with large winter crops should not generally have any problems meeting the 65% green field requirement in the autumn period. But in many other cases it will be necessary to sow second crops, in most cases under a main crop in the spring. The problem is that grain is harvested relatively late in Denmark, often well into September, so that it is not possible to sow second crops which will succeed in growing for the autumn period. By means of sample surveys the authorities will check to ensure that fertilization programmes have been established for farms and that the area of green fields, i.e. fields bearing growth in the autumn period, are respected.

Advisory services have always warned against exaggerated use of artificial fertilizers (primarily nitrogen) on economic, environmental and fertilizer quality grounds. The establishment of the KVADRATNET survey for nitrates in Denmark makes it possible to make more detailed N-forecasts related to regional conditions soil type and type of farming.

The Danish environmental authorities have focussed closely on nitrogen run-off as the main cause of increasing eutrofication of lakes, watercourses and the sea. Politicians have also adopted this approach.

There is recognition now that phosphorus, rather than nitrogen, is a greater factor in the pollution of lakes and coastal areas in particular. It is the hope of the agricultural sector that it will be possible, to persuade the public authorities that a reduction in the application of nitrogen from chemical fertilizers to below the economically optimal levels will not reduce nitrogen run-off to any noticeable extent but will simply reduce profitability and agricultural earnings. Farmers have pointed out that there are other factors with regard to run-off which are more significant, e.g. choice and rotation of crops, winter cropping, soil treatment and especially the use of farmyard manure.

In the last few years it has been suggested in Parliament that a duty or tax be applied to nitrogen from chemical fertilizers with the aim of reducing the optimal economic quantity and thereby reducing nitrogen use in farming. The proposal was not adopted but is often used as a threat if farmers do not reduce nitrogen levels voluntarily. The problem with using a tax on nitrogen to control nitrogen use is that the tax would need to be very high - at least 2 to 3 times the current price of nitrogen - if it is to have an effect on nitrogen consumption. As well as leading to a worsening in farm profitability, it has been accepted that this proposal has a fundamental weakness in that a reduction in consumption below the economically optimal level would have no measurable effect on nitrate leaching. A high price of nitrogen would also lead to increased use of pulses in crop rotation, which would not be conducive to solving the leaching problem.

Animal Manures

The use of animal manure is regulated by the Agricultural Notice No. 668 of October 14, 1987. The main features are as follows:
1. After a transitional period, storage capacity for farmyard manure shall generally be sufficient for 9 months.
2. Herd density may not exceed the following limits:
    Cattle farming; 2.3 Animal units (DE) per ha.
    Pig farming; 1.7 DE per ha.
    Arable farming; 1.7 DE per ha.

If the animal population exceeds these limits, agreements on the disposal of manure must be made with neighbouring establishments.

3. For liquid manures the following rules apply:
   - Manure spread on bare fields must be ploughed-in within 12 hours of application.
   - Manure may not be spread on areas without vegetation which are frozen or snow-covered.
   - Spreading is not permitted on areas without vegetation between harvest and November 1.
   - Spreading may not take place at weekends closer than 200m from residential areas.
   - There must be no run-off into watercourses etc.
4. Local authorities may add special rules to the above if they consider that the use of farmyard manure is giving rise to nuisance.

The main problem experienced with these regulations is the cost of investment in increased storage capacity. Another practical problem is that some of the farms which are obliged by these regulations to transport their animal manure to other livestock-free establishments have difficulty finding outlets for the surplus manure. Farmyard manure is not attractive to farmers growing cash crops because it cannot be divided up and applied with the same precision as chemical fertilizer.

In general, it is believed that the most effective arrangement is that which limits animal density per hectare. The rules regarding intensive livestock farming, storage and use of farmyard manure which come into force in the next few years will mean that a significant number of farmers will be expected to withdraw from livestock farming. This has become a matter of national concern. For this reason, there are discussions taking place to see whether some of the regulations introduced, including the 9-month storage capacity requirement, are not too restrictive.

In Denmark the main danger of pollution from intensive farming is felt to be connected with the use of farmyard manure. It is estimated that on average, there is not over-use of nitrogen from chemical fertilizers.

Problems arise because the areas under grass, both permanently and in the course of crop rotation, are relatively small and decreasing. Furthermore many one-year crops are grown because over-wintering possibilities for crops are generally not good under Danish climatic conditions. The combination of better use of farmyard manure, spreading over a larger area and at more appropriate times, together with a greater proportion of land under vegetation during the autumn will certainly mean less seepage of nitrate leaching from arable land. The loss of phosphorus in farming seems relatively unimportant, but this question must be examined more closely and possible phosphorus loss must be limited as this nutrient is increasingly being viewed as a significant factor in open water pollution.

FEDERAL REPUBLIC OF GERMANY

Chemical fertilizers

The law on chemical fertilizers does not directly concern agricultural activities. It only regulates trade of fertilizers. According to the Fertilizer Law, fertilizers can be marketed only if they meet a standard and take into account the objectives of increasing growth, yield, and quality of useful plants or the production capacity of the soil. The application of fertilizers by farmers is not affected by this law.

Animal manures

The spreading of slurry, manure, and stable manure can be regulated by the Law on Waste Disposal. Farmyard manures are excluded from the category of waste once they are used to fertilise farmland. Therefore, the matter is treated on a case-by-case basis.

After long years of discussion, the Federal Government rejected the idea of national regulations and has left it up to the Länder to develop appropriate laws. Some Länder have developed regulations (Nordrhein-Westfalen has the Manure Ordinance and Niedersachsen has a Manure Decree). In both regulations, there is a prohibition of spreading manure in the winter and a limitation of maximum permissible amounts of fertilizer to 240 kg N per ha per year.

Restrictions under other regulations

In the interest of water protection, agricultural land use and fertilization in protected water areas may be subject to further restrictions by the Law on Water Resources Management. As regards environmental protection, farming can also be limited - particularly in protected sectors of nature and the countryside by the Federal Natural Protection Law (BNatSchg). This can affect not only chemical fertilizers, but also the use of farmyard manure. Restrictions would be aimed at certain protection goals (i.e., maintenance of plants in nutrient-poor locations).

The Food and Basic Consumer Goods Law has an indirect influence on agricultural fertilization. On the one hand, it can set a maximum ceiling which regulates the amount of nitrate in food. For a long time, the enabling act of this law has been invoked only in relation to baby food. In addition, the Drink Water Ordinance, which has been transferred from a corresponding EC regulation into national law, is based on provisions of the Food Law. The setting of a threshold value for nitrate in drinking water has an indirect influence on the setting up of protected water areas.

The main problems encountered in applying the above regulations include:-

1. The Manure Ordinance and Manure Decree hinder only extreme cases of pollution from spreading of manure. Because of the time limitation on manure spreading for farms with high livestock density, nitrate leaching problems can worsen. It would be more appropriate to create regulations based on specific local conditions but this may be hampered by administrative difficulties.
2. Limited manure storage facilities in certain areas where manure spreading is confined to specific time periods. In some Länder, there are comprehensive measures taken to expand the needed storage capacity.
3. With regard to the provisions in the protection water areas of Baden-Württemberg, there are implementation problems in carrying out and analyzing soil samples. The setting of a threshold value for nitrogen in the soil is probably not justifiable because of weather, soil conditions, crop rotation, and soil cultivation practices all have more influence than nitrogen application on nitrate levels.
4. In general, it seems that the verification of fertilization restrictions is very difficult and is linked to considerable administrative costs.

There are problems in evaluating organic fertilizers with regard to their nutrient content and delivery of nutrients and this is a research priority. This is particularly the case with processing farms, which show a considerable excess of nitrogen that can be traced to the proportion of fodder purchased.

Farmers should be able to get around problems related to ploughing up large areas of pasture land. Through intensive advisory efforts and education, the farmer should be encouraged to plant catch crops so that the danger of nutrient leaching in the winter can be greatly reduced. In addition, there are several measures to increase production, such as plant protection adapted to local conditions, cultivation of higher yield strains, spray irrigation in dry areas, etc., and when the level of nitrogen fertilization remains the same, to improve the use of nitrogen fertilizer and, thereby, reduce the danger of nitrate leaching.

Solutions are also being sought for horticulture, especially on light soils, which will take the mineralised nitrogen in the soil into consideration. Thought is also being given to other horticultural aspects, such as the cultivation of catch crops and better crop rotation. For this reason, the amount of fertilization in special crops such as asparagus, winestock and fruit crops has recently declined in Germany.

Through the development of nitrification inhibitors, the nitrification of ammonia (in mineral fertilizers or organic fertilizers) is slowed down for a certain amount of time. Therefore, the danger of nitrate displacement is reduced, especially in crops with slow early growth (particularly corn, but also sugar beet and potatoes).

In farming, the environmental problems must be tackled and a "Code of good agricultural practice" developed. In the interest of the water supply, restrictions on agriculture may be necessary, which in certain areas may limit good agricultural practice and thus raise the question of compensation to farmers.

Research, development, and advisory services should be more strongly promoted, so that the farmer will learn about production procedures which are both more economical and more environmentally sound. Existing procedures should be used consistently and promoted more in the future.

To counterbalance the administrative costs entailed in increased inspection and monitoring for compliance, farmers should be educated about the need for more efficient means of water protection.

GREECE

Chemical Fertilizers

In relation to chemical fertilizers, there are no specific laws, regulations or administrative provisions at this time. The Ministry of Agriculture organizes a series of educational seminars which recommend amounts of chemical fertilizers for application to crops. It also encourages the use of organic fertilizers such as animal and poultry wastes in a controlled manner. In addition, five or six monitoring stations for chemical fertilizers use are shortly to be established by the Ministry.

Animal Manures

The following regulations currently apply with regard to the management, treatment and disposal of livestock and poultry wastes.

1. Regulation No. 8181/87 of the Ministry of Public Health deals with conditions for installation and proper operation of animal farms. It also covers rules and measures for proper management and treatment of solid and liquid wastes prior to their disposal. Concerning liquid wastes, along with some basic guidelines for good management and disposal, it refers to a more general regulation of the Ministry of Public Health (No. E1b221/65) for their treatment. Both aerobic and/or anaerobic biological treatment of liquid wastes is required, as well as some mechanical separation of suspended solids from the liquid phase of wastes. Very strict limits, BOD_5 less than 50 mg/litre and waste less than 1200 mg/litre are imposed for liquid waste disposal in water bodies and on land respectively.
There is no specific requirement for nitrogen or other mineral removal prior to disposal, except that of preferring land disposal for direct disposal of liquid wastes to water.
2. Regulations No. 83840/3591/87 of the Ministry of Environment and Public Works deals with the proper distance of animal farms from cities, communities, villages, national or main roads, railways, beaches, schools, hospitals etc. The purpose is to eliminate

problems resulting from odours and generally the existence of these farms close to residential areas.

The main problems in applying these regulations are:
- Difficulties with the proper control of the measures required by each regulation especially that of the biological treatment of wastes;
- Strict water quality standards in the case of large intensive livestock farms can lead to uneconomical solutions for waste management and treatment;
- Lack of training on environment protection and awareness of its importance make it difficult in many cases, to promote understanding of the importance of such regulations.

Other than measures related to the disposal of animal liquid wastes in a controlled manner and requiring an efficient and economical treatment of wastes, no specific innovations for protection of water quality are expected for the time being. The future orientation of policy with regard to agriculture and the environment will be concerned with:

1. The integration of livestock farms in plant production systems which recycle wastes to the land and reduce the need for chemical fertilizers;
2. Education and participation of farmers in attempts to protect the environment in order to make the application of laws and regulations easier and more efficient;
3. The expansion of livestock and poultry farms in number and size in accordance with the requirements for protection of the environment.

SPAIN

Chemical fertilizers

No national laws exist which control this area of chemical fertilizers by farmers. This is also the situation also in the autonomous regions of Spain.

Animal manures

No national or regional laws exist in Spain, specifically related to monitoring and regulating intensive livestock farming and the use of animal manures.

Intensive farming continues to be regulated by general legislation. National provisions include:

(1) The Regulation on Annoying, Unhealthy, Harmful and Dangerous Activities, Decree 2414/1961 of November 30. The regulation covers the following agricultural activities:
- dairy farming
- pig farming
- bird raising
- rabbit breeding
- obtaining organic fertilizers
- waste disposal through auto-purification
- waste disposal through biological purification
- waste disposal through physics (sic) techniques
- waste disposal through biological means

Location of these activities is governed by Municipal Orders and Urban Planning laws. If none exists, the decision is up to the province's Technical Services Committee, as are the corresponding corrective measures.

When the commencement of these classified activities presents a danger of water pollution, the regulation refers back to compliance with the Water Law described below.

Art. 17 of the Regulation lists conditions where manuring can occur and the characteristics of the slurry to be used.

The main problem with the Regulation on annoying, unhealthy, harmful and dangerous activities is that many of the districts concerned do not have adequate management and monitoring capacity, particularly in rural districts.

2. Water Law of August 2, 1985 (BOE no. 1891) which became effective on January 1, 1986 and provides for, inter alia:-
    a) establishment of the unit for water management, whether it be surface or underground supplies
    b) establishment of water supply organizations, which group the former Confederations and Commissariats. The latter have been in operation since 1958. The main functions of the new organizations are to administer and monitor public water supplies; plan, build and harness their own works and those it builds for the State; monitor water quality and define quality objectives and programmes.
    c) establishment of a general procedure for avoiding contamination. The procedure states that all manuring requires prior authorization which may be revoked if conditions are not met. It also establishes a levy for the use of public water supplies. All authorised manuring is also taxed with a levy to protect and improve the area affected.

A manuring levy is a new concept in Spain and is applied according to the following criteria: All sewage will have a certain minimum quality before being spread; and contamination units (cu's) are set according to the rated quality of manures. A formula will be applied to cu's to determine how much must be paid to the water supply organization. Revenue will go towards protecting water sources.
    d) Infractions and Fines
        In addition to the above procedure, fines of up to 50 million

ptas. are charged. The offender may also have to repair damage to the water supply and return it to its former state. These fines can be much higher than those charged previously.

3) The Air Protection Law of December 1972 classifies as air polluting activities: stables with over 100 head of cattle, barnyards with over 10 000 birds. It requires a series of satisfied prior conditions to enable construction of these holdings, including a study of environmental impact and ways to repair damage.

4) Royal Legislative Decree 1302/1986 of June 28 on Environmental Impact, transposing EC Directive 85/337. Although this Decree does not state that intensive farming must be subject to an Environmental Impact Evaluation, Article 2 of the Directive refers to projects "likely to have a considerable effect on the environment due to their nature, size or location" and these activities definitely have a considerable effect on the environment in many cases.

Not enough time has gone by to assess the Water Law since it is still in the organizational stages. However, there are problems with applying the manuring levy because the situation has gone from one of relative tolerance (even though adequate laws existed) to one of demanding payment and applying large fines for non-compliance. The Decree on Environmental Impact Evaluation does not come into force until June of 1988.

The political will now exists to monitor the quality of both surface and ground water supplies. Intensive livestock and crop farming are not priority environmental concerns since they affect proportionally small areas of the whole country. Much of Spanish soil is very low in organic matter and is therefore able to assimilate a great deal of organic slurry. Of great concern is the excessive tapping of water supplies on the Mediterranean seaboard and the resulting infiltration of seawater and other ecological and economic repercussions.

FRANCE

Chemical Fertilizers

The main laws, regulations and administrative provisions in France in this area are standardization to protect the user (farmers and non-farmers), and confirmation to ensure that chemical fertilizers are harmless and effective. If it is confirmed, a temporary sales permit is issued. Specific measures can be taken in certain areas to protect water courses. If imposed fertilization limits cause "direct, material and definite damage", then farmers can be compensated.

An action programme has been adopted to help in the preventive struggle against nitrate water pollution caused by farming. Its main purpose is to improve farm practice in order to reduce nitrate leaching and to provide complete information on this topic to agricultural advisors and farmers. Emphasis is placed on promoting rational fertilizing which means using only the necessary amount of nitrogen on crops and at the right time.

It is considered impossible to define the notion of "abusive use of fertilizers" given the current state of knowledge in this regard. France's well-known geographical diversity and particularly the varying soil and climatic conditions require accurate data before the term can be defined for different farming regions. However, progress can be expected regarding leaching and run-off risks by improving soil coverage (green fertilizers) and turning under of crop debris.

Animal Manures

Specific provisions relating to livestock farming come from three sources:
- Provisions governing water, particularly the law of 1964,
- Department of Health regulation (1983),
- The law concerning classified facilities (1976).

These laws are broad in scope and livestock farming is one of the activities covered. Large livestock farms are subject to stricter rules than small ones:

- Small livestock farms must follow a declaration procedure in order to be set up or enlarged.
- All large livestock farms, except for dairy cattle, and sheep farms, must be authorized. Regulations govern the animal/crop balance and manure spreading. Farms are monitored in accordance with these regulations.

Slurry is a problem mainly in pig and poultry farming, but much less so in cattle farming. It is governed by specific rules (manure spreading plan). Farmers cannot spread manure within 200 m of neighbouring homes clearly reducing the land available for animal manure.

France has chosen to favour preventive measures in the area of nitrate levels in water. Most of the problems caused by all types of animal farming can be solved by cropping appropriate to the environments. Considering all the technical problems as a first step makes it easier to contemplate possible legal measures later.

The Ministers for Agriculture and for the Environment adopted an action programme in October 1987 to ease the problems of intensive farming. The programme refers to the following areas:

1. Quality of excreta (e.g. Avoiding dilution)
2. Storage: Nationally the minimum legal storage capacity for slurry is 45 days. The ideal limit in Brittany would be about six months, because of rain conditions. The target in practice should be a minimum of 4 months. For manure, the goal is better collection of liquid and solid manure, from uncovered work areas.
3. Manuring:
    - Better combination of chemical fertilizers and livestock manure.
    - Better manuring through manuring warnings
    - Better monitoring of manuring plans
    - Better techniques and equipment
4. Transport and/or Processing:
    - Advertisements supply and demand for slurry and other excreta through minitel.

- Study processing routes - either moist (compost) or dry (granulation) routes - for poultry excreta with a view to transportation.
- Study pig slurry processing routes.

IRELAND

Chemical Fertilizers

There are no regulations concerning the use and monitoring of chemical fertilizers in Ireland at the present.

Animal Manures

There is no specific provisions controlling the application of animal manures on land under the local government (Water Pollution) Act, 1977 and the various EC directives.

A farmer, like anybody else, is subject to Section 3 of the 1977 Act which provides that "a person shall not cause or permit any polluting matter to enter waters" An acceptable defence, for a person charged with such an offence, is to prove that (s)he took all reasonable care to prevent the prohibited entry. "All reasonable care" is judged in the light of current good agricultural practice, regarding agricultural effluent. Under Section 12 of the 1977 Act local authorities are empowered to require specific steps to be taken to prevent polluting matter entering waters from premises (eg silos, livestock housing, slurry tanks, dungsteads).

Control under the act has been weak because monitoring for compliance and enforcement has been minimal. Also the fines laid down in the legislation are small and hardly prohibitive. Its probably true to say that environmental constraints were regarded as a nuisance by those engaged in agricultural development.

The scene is changing, particularly in respect of new agricultural building developments. Since 1984 (Statutory Instrument 348 of 1984) planning permission must be obtained for the erection of:
- Roofed livestock housing, including effluent storage,
- Roofless facilities, silage areas etc, and
- Farm buildings not for livestock

where the aggregate floor area for each class of structure exceeds 300 sq metres. Regulations made in 1977 (Statutory Instrument 65 of 1977) exempted structures of less than 400 sq metres from planning control. Local authorities insist on certain environmental guarantees appropriate to the specific planning application. These guarantees would generally be stricter where planning decisions are made subsequent to objections raised leading to an appeal to the planning board (An Bord Pleanala).

The main controls on agriculture have been extra-legal until recently. This was usually achieved through conditions attached for grant aid under the farm modernization scheme, derived in the case of water pollution control, from Department of Agriculture guidelines and recommendations. The Department of Agriculture recommends:
- Manure storage capacity of 16 to 26 weeks
- Non-spreading of manure within 12 to 40 metres of a river or stream (40 metres applies more to lakes).
- Access to 28 hectares of suitable land per 1000 pig places.

In certain planning cases additional and tighter constraints have been imposed (e.g. a ban on manure spreading over certain months, dry feed only to pigs) usually but not exclusively, in the case of pig units. Generally there is not a problem of over-application of cattle manures on available land.

The government Programme on Water Pollution (1987) aims to "strengthen existing legislation in the area and it will make sure that the polluter pays for the damage done". The programme includes:
- A survey of farms to identify potential sources of water pollution, with follow-up action to be taken under the 1977 Act where pollution prevention measures are required, or under the Planning Acts in the case of unauthorized developments.
- Awareness/education programmes.
- Further restricting agricultural development which can be undertaken without planning permission.
- Heavier penalties under 1977 act.
- Prohibition of certain agricultural practices by local authorities where considered appropriate.

- A new water pollution bill with possible powers enabling local authorities to require farmers to register if they wish to carry out certain activities (eg slurry spreading) in areas where there is a risk of water pollution. The local authorities could prohibit some activities or require that certain conditions are adhered to.

ITALY

Existing legislation sets no constraints on the use of chemical fertilizers.

No reference was made in the SEPFAR submission to controls on animal manure disposal or livestock intensity.

LUXEMBOURG

In general, national and district administrations prefer to establish zones of protection around the water springs tapped, as well as to restrict or prohibit activities that may cause underground or surface water pollution within these zones. These zones are delineated according to the commonly used system of: (1) Catchment Zones, (11) Immediate Protection, (111) and Wide Protection Zones. Article 3 of the Law of June 27, 1906 on public health protection has long provided a possibility for planning protection zones around sources of drinking water. However, there has been little concrete action towards actually setting up such protection zones. Other laws for the protection of surface and underground water, (e.g. Law of May 29, 1929 on the clearing, maintenance, and improvement of waterways and the Law of January 9, 1961 on the protection of underground water) have had little impact. Up to the present time, 19 districts have considered setting up protection zones within their boundaries. A model regulation is made available to them by the Environment Administration.

For the water supply authorities, the absence of legislation adapted to their specific situation poses administrative and political problems. Draft legislation law is being prepared for the purpose of fighting water pollution.

Environmental problems related to intensive farming - particularly of pigs - are addressed in the Grand Ducal Regulation of March 18, 1982. Among other things, the Regulation specifies conditions for storage and spreading of solid and liquid manure from pig units.

In order to protect the quality of the water from the Lac de La Haute-Sure, special measures have been implemented. The Law of May 27, 1961 on health protection for the Esch-sur-Sure dam was the basis for protective measures and created a protection zone totaling 44.11 sq km.

This zone includes the lake shores and is subdivided into Zone I (978 hectares) and Zone II (3,433 hectares). The Grand Ducal Regulation of 7 March 1985 assures the protection of the water in Zone II. The project for planning the "Haute-Sure Natural Park" has elements specifically for the protection of the lake as a reservoir for drinking water.

A draft, which was presented in 1986 by the Ministry of State and the Ministry of Land Planning, provides a number of measures, one of which is the setting up of a supplementary protection zone (Zone III) of 7,900 hectares. The legal framework of this draft includes:
- The Law of March 20, 1974 on general land planning
- The Law of August 11, 1982 on protection of the environment and of natural resources.

Due to fierce opposition, from both the farmers and the districts in the region, the implementation of the project in question was postponed. Opponents, of the project in its present state felt that it was not feasible. At the same time the Luxembourg Water Services Authority (ALUSEAU), have since requested new legislation requiring districts to set up protection zones as part of an integrated land planning policy.
Recommendations from Ministries of Agriculture and the Environment regarding the spreading of liquid manure are also in existence.

NETHERLANDS

Chemical Fertilizers

There are no legally enforceable rules and standards in the Netherlands governing the use of chemical fertilizers. There is at present a draft regulation forbidding the use of chemical fertilizers along with animal manures in water collection areas - i.e. areas in the immediate vicinity of boreholes.

Animal Manures

The legal protection of the groundwater is only just getting under way, with attention focussed initially on animal manure. On January 1, 1987 the new Soil Protection Act ("Wet Bodembescherming") came into effect, and includes new provisions for the protection of groundwater destined for human consumption. Prior to January 1, 1987, the groundwater had been protected by a range of administrative provisions. As the protection of groundwater was not the primary target of these regulations, or their regulatory power was very weak, they proved to be inadequate in practice.

The Soil Protection Act is a general or outline law. In other words, it does not itself lay down limits or rules, but rather authorizes the government to introduce these via General Administrative Measures ("Algemene Maatregelen van Bestuur"). The introduction of national regulations within the framework of this law will provide a general level of protection. One example of such protection is the GAM entitled "Order relating to the Use of Animal Fertilizers", setting forth rules on how much animal manure can be spread at what time of year and by what methods.

Maximum levels for the application of animal manure are based on its phosphate (P_2O_5) content. These new standards are being phased in gradually to enable farmers to adapt their farming methods. Manure spreading is prohibited at certain times:

Grasslands: October 1 to December 1,
January 1 to February 15 if ground is snow covered.

Sandy cropland: From harvest to November 1,
October 1 to November 1 if an after-crop is cultivated.

Also manure spread on cropland must be worked into the ground no later than 1 day following spreading. The phosphate content limits came into effect on May 1, 1987, whilst the rules governing manure spreading became effective from January 1, 1988.

In addition to national protection levels the provinces are required to take additional measures to protect groundwater intended for use as drinking water. The Soil Protection Act requires every Province to draw up a Groundwater Protection Plan (GPP) and Groundwater Protection Bye-laws (GPB), with the GPP providing the main lines of the groundwater protection policy as well as setting forth the size and location of the groundwater protection zones. In the GPB the proposed policies are converted into regulations for preventing the contamination of the groundwater. Draft GPP's and GPB's have already been completed in various provinces in the sandy zones. These Bye-laws have the effect of stiffening national regulations governing the use of animal manure. The standards may be adjusted over time, the final goal being to attain a target of max. 25 mg NO_3/litre in groundwater intended for use as drinking water. The vulnerable groundwater protection zones are designated by the provincial authorities.

In addition, the application of the Town and Country Planning Policy ("Ruimtelijk Ordeningsbeleid") will have the effect of curbing the establishment and extension of intensive animal farming in the groundwater protection areas. Under the terms of the Fertilizers Act ("Meststoffenwet") every farming unit is allowed to spread up to 125 Kg P_2O_5 per ha of animal manure. Compliance will be monitored by use of "Fertilizer books". Every farm producing more than 125 kg P_2O_5/ha/year must maintain a "Fertilizer book" containing details of the number of animals, related fertilizer production and the disposal of the surplus outside the farm unit, and which can be inspected by the General Inspectorate ("Algemene Ispectiedienst").

Two major problems arising from these regulations are the fertilizer surpluses resulting from the phosphate norms and monitoring the application of these standards. The practicability and effectiveness of the new regulations will be governed by the extent to which these problems can be solved. The fertilizer bookkeeping will help to determine if the excess fertilizer has indeed been disposed of on or off the farm. On the other hand, it is much more difficult to monitor on which fields the manure has been spread. If necessary, soil sampling will be undertaken if there are clear indications that the permitted levels have been exceeded (so-called manure dumping). Since these additional provisions have not yet come into effect, there is as yet little insight as to the success or otherwise of the inspection of fertilizer books by the General Inspectorate. The final shape of monitoring in the groundwater protection areas still remains a matter for discussion.

In the initial phase the national phosphate contents limits will produce a surplus of some 14 million tons of animal manure on the farms concerned. The phosphate limits have been drawn up in such a way that this excess can be disposed of in other parts of the country. (i.e. no national surplus). The tightening up of the limits in phase two will give rise to an overall national surplus of approx. five million tons by 1991. Problems occurred in autumn 1987 with fertilizer disposal due to the weather conditions. Protective measures in groundwater protection areas will lead to further manure surpluses, with these surpluses being most serious in sandy areas. The introduction of these regulations will require large-scale industrial processing of animal manures. Technical solutions for the processing of animal fertilizer are still at the development stage. For this reason solutions are being looked for elsewhere; these include increased disposal of animal manure in crop growing areas, a well-organized distribution system, improving the quality of manure from livestock units and the reduced use of minerals in feed concentrates.

PORTUGAL

Currently there is no legislation related to the use of chemical fertilizers by farmers. Similarly, no legislative controls exist in the use of animal Manures.

Concern about the deleterious effects of intensive farming practice is growing particularly in relation to intensive chemical spraying of cereals in certain regions. Administrative structures are not as yet in place to deal adequately with problems arising out of intensive agriculture where they occur.

The use of animal manures is authorised by the Pollution Control Services where the farmer has access to facilities for treatment and purification of effluent and sufficient own land area for disposal of manures. Lines of credit for purification facilities are available to farmers and the importance of farmer education and advice on pollution is recognised while recent and impending legislation is aimed at securing greater environmental protection in general.

UNITED KINGDOM

Chemical Fertilizers

The Control of Pollution Act (1974) is the main legal instrument for controlling pollution of the environment, including water. Under this Act, the Minister for Agriculture issues the Code of Good Agricultural Practice and it is updated according to the latest Advisory Service practice. The act states that:

> "Application rates of fertilizers should take account of crop requirements and the nutrients provided by any organic manures and the soil. To reduce the danger of nutrients being leached out and polluting relevant waters, fertilizers (particularly nitrogenous fertilizers) should not exceed maximum ADAS recommended rates.
>
> Nitrogenous fertilizers should only be applied at times when the crops can utilise the nitrogen. In autumn and winter application should be avoided except when there is a specific crop requirement. During the autumn, nitrate is produced in the soil from organic nitrogen. Cultivation of the soil tends to increase the amount of nitrate produced. Seed bed applications of nitrogen for winter cereals are therefore rarely necessary, except for direct drilled crops."

The UK Government is currently considering the imposition of controls on agriculture in specific catchments. This could give the Secretary of State for the Environment powers to prohibit or restrict activities in a particular area with a view to protecting surface or ground water from pollution. These powers have not hitherto been tested. The UK is committed to meet EC requirements but achieving these within the next few years will not be easy in many cases.

Animal Manures

The Code of Good Agricultural Practice lays down procedures to minimize the risks to pollution of water. Manures should be applied:

> "to avoid direct contamination of relevant waters. They should not be applied in the vicinity of a groundwater supply source

> or within a zone of protection as notified by the Water Authority."

Also:
> "Gradient, soil type, degree of saturation or desiccation, temperature and vegetation will all have effects on the extent to which applied manures are absorbed or run off. When conditions increase the risk of pollution, spreading should be undertaken only if all reasonable precautions are taken to avoid such pollution and spreading should cease if pollution is obviously going to occur."

Although the control of odour from livestock units is not directly connected with water pollution, the legal controls on odour emission under the Public Health Act (1936), the Public Health (Recurrent Nuisances) Act (1969), the Town and Country Planning Act (1957) and its dependent General Development Orders have implications for water pollution. Local Authorities (not Water Authorities) have power in certain circumstances, to consider whether the initiation or enlargement of an animal holding should be permitted. The UK Government is consulting on possible amendments to the relevant General Development Order which may result in closer control over developments in animal holdings.

Another form of control arises from the operation of the grant system to farmers under EC rules. For fixed equipment related to the storage and treatment of manures, particularly high grants (30% generally and 60% in less favoured areas) are possible. For these to be paid the farmer must show that the Water Authority has approved his proposal. These grants relate to slurry and manure stores, silage effluent and yard washings which occasion much of the pollution. Their introduction in 1985 was intended to help reverse the upward trend of pollution incidents.

Some of the major problems seen with this proposal are:
1. The cost to farmers of manure storage and handling facilities
2. The need for education of farmers regarding the use of manures and the risks of pollution, and

3. The lack of data and understanding on which to base regulations for "good farming practice".

ission

t on the environment and the rural economy of
cal and animal fertilizers

blications of the European Communities

EN

ISBN 92-826-0123-4

Catalogue number: CB-57-89-386-EN-C

Price (excluding VAT) in Luxembourg: ECU 20.25

Venta y suscripciones · Salg og abonnement · Verkauf und Abonnement · Πωλήσεις και συνδρομές
Sales and subscriptions · Vente et abonnements · Vendita e abbonamenti
Verkoop en abonnementen · Venda e assinaturas

BELGIQUE / BELGIË

Moniteur belge / Belgisch Staatsblad
42, Rue de Louvain / Leuvenseweg 42
1000 Bruxelles / 1000 Brussel
Tél. 512 00 26
Télécopieur: 511 01 84
CCP / Postrekening 000-2005502-27

Sous-dépôts / Agentschappen:

Librairie européenne / Europese Boekhandel
Avenue Albert Jonnart 50 / Albert Jonnartlaan 50
1200 Bruxelles / 1200 Brussel
Tél. 734 02 81
Télécopieur: 735 08 60

Jean De Lannoy
Avenue du Roi 202 /Koningslaan 202
1060 Bruxelles / 1060 Brussel
Tél. (02) 538 5169
Télex 63220 UNBOOK B

CREDOC
Rue de la Montagne 34 / Bergstraat 34
Bte 11 / Bus 11
1000 Bruxelles / 1000 Brussel

DANMARK

J. H. Schultz Information A/S
EF-Publikationer
Ottiliavej 18
2500 Valby
Tlf: 36 44 22 66
Telefax: 36 44 01 41
Girokonto 6 00 08 86

BR DEUTSCHLAND

Bundesanzeiger Verlag
Breite Straße
Postfach 10 80 06
5000 Köln 1
Tel. (02 21) 20 29-0
Fernschreiber:
ANZEIGER BONN 8 882 595
Telefax: 20 29 278

GREECE

G.C. Eleftheroudakis SA
International Bookstore
4 Nikis Street
105 63 Athens
Tel. 3226-323
Telex: 219410 ELEF
Telefax: 3254 889

Sub-agent for Northern Greece:

Molho's Bookstore
The Business Bookshop
10 Tsimiski Street
Thessaloniki
Tel. 275 271
Telex 412885 LIMO

ESPAÑA

Boletín Oficial del Estado
Trafalgar 27
E-28010 Madrid
Tel. (91) 446 60 00

Mundi-Prensa Libros, S.A.
Castelló 37
E-28001 Madrid
Tel. (91) 431 33 99 (Libros)
     431 32 22 (Suscripciones)
     435 36 37 (Dirección)
Télex 49370-MPLI-E
Telefax: (91) 275 39 98

FRANCE

Journal officiel
Service des publications
des Communautés européennes
26, rue Desaix
75727 Paris Cedex 15
Tél. (1) 40 58 75 00
Télécopieur: (1) 4058 7574

IRELAND

Government Publications Sales Office
Sun Alliance House
Molesworth Street
Dublin 2
Tel. 71 03 09

or by post

Government Stationery Office
EEC Section
6th floor
Bishop Street
Dublin 8
Tel. 78 16 66

ITALIA

Licosa Spa
Via Benedetto Fortini, 120/10
Casella postale 552
50 125 Firenze
Tel. 64 54 15
Telefax: 64 12 57
Telex 570466 LICOSA I
CCP 343 509

Subagenti:

Libreria scientifica Lucio de Biasio - AEIOU
Via Meravigli, 16
20 123 Milano
Tel. 80 76 79

Herder Editrice e Libreria
Piazza Montecitorio, 117-120
00 186 Roma
Tel. 67 94 628/67 95 304

Libreria giuridica
Via 12 Ottobre, 172/R
16 121 Genova
Tel. 59 56 93

GRAND-DUCHÉ DE LUXEMBOURG

Abonnements seulement
Subscriptions only
Nur für Abonnements

Messageries Paul Kraus
11, rue Christophe Plantin
L-2339 Luxembourg
Tél. 48 21 31
Télex 2615
CCP 49242-63

NEDERLAND

SDU uitgeverij
Christoffel Plantijnstraat 2
Postbus 20014
2500 EA 's-Gravenhage
Tel. (070) 78 98 80 (bestellingen)
Telefax: (070) 476351

PORTUGAL

Imprensa Nacional
Casa da Moeda, E.P.
Rua D. Francisco Manuel de Melo, 5
1092 Lisboa Codex
Tel. 69 34 14

Distribuidora Livros Bertrand Lda.
Grupo Bertrand, SARL
Rua das Terras dos Vales, 4-A
Apart. 37
2700 Amadora Codex
Tel. 493 90 50 - 494 87 88
Telex 15798 BERDIS

UNITED KINGDOM

HMSO Books (PC 16)
HMSO Publications Centre
51 Nine Elms Lane
London SW8 5DR
Tel. (01) 873 9090
Fax: GP3 873 8463

Sub-agent:

Alan Armstrong Ltd
2 Arkwright Road
Reading, Berks RG2 0SQ
Tel. (0734) 75 17 71
Telex 849937 AAALTD G
Fax: (0734) 755164

SUISSE

OSEC
Stampfenbachstraße 85
CH-8035 Zurich
Tél. (01) 365 51 51
Fax: (01) 365 52 21

ÖSTERREICH

Manz'sche Verlagsbuchhandlung
Kohlmarkt 16
1014 Wien
Tel. (0222) 531 61-0
Telex 11 25 00 BOX A
Telefax: (0222) 531 61-81

TÜRKIYE

Dünya süper veb ofset A.Ş.
Narlıbahçe Sokak No. 15
Cağaloğlu
Istanbul
Tel. 512 01 90
Telex: 23822 dsvo-tr

UNITED STATES OF AMERICA

UNIPUB
4661-F Assembly Drive
Lanham, MD 20706-4391
toll free (800) 274-4888
Fax: (301) 459-0056
Télex 7108260418

CANADA

Renouf Publishing Co., Ltd
61 Sparks Street
Ottawa
Ontario K1P 5R1
Tel. Toll Free 1 (800) 267 4164
Ottawa Region (613) 238 8985-6
Telex 053-4936

JAPAN

Kinokuniya Company Ltd
17-7 Shinjuku 3-Chome
Shinjuku-ku
Tokyo 160-91
Tel. (03) 354 0131

Journal Department
PO Box 55 Chitose
Tokyo 156
Tel. (03) 439 0124

AUTRES PAYS
OTHER COUNTRIES
ANDERE LÄNDER

**Office des publications officielles
des Communautés européennes**
2, rue Mercier
L-2985 Luxembourg
Tél. 49 92 81
Télex PUBOF LU 1324 b
CC bancaire BIL 8-109/6003/700

Catalogue number: CB-57-89-386-EN-C

Price (excluding VAT) in Luxembourg: ECU 20.25

OFFICE FOR OFFICIAL PUBLICATIONS
OF THE EUROPEAN COMMUNITIES
L-2985 Luxembourg

ISBN 92-826-0123-4